Human Evolution

Adolph H. Schultz

Human Evolution

Biosocial Perspectives

Edited by S. L. Washburn and Elizabeth R. McCown

Perspectives on Human Evolution, Volume IV
A Publication of the Society for the Study of Human Evolution, Inc., Berkeley, California

The Benjamin/Cummings Publishing Company
Menlo Park, California • Reading, Massachusetts
London • Amsterdam • Don Mills, Ontario • Sydney

This book is in the
Benjamin/Cummings Series in Anthropology

Consulting Editor
Brian Fagan

ISBN–0–8053–9517–2
ABCDEFGHIJ–HA–798

The Benjamin/Cummings Publishing Company
2727 Sand Hill Road
Menlo Park, California 94025

Human Evolution
Biosocial Perspectives

is a book sponsored by:

Society for the Study of Human Evolution, Incorporated
Berkeley, California

Organized for the purpose of increasing and disseminating information in the field of human evolution through a series of books entitled Perspectives on Human Evolution.

Officers of the Society for 1977:
Vincent M. Sarich, President; Phyllis Dolhinow,
S. L. Washburn, Alice Davis, Elizabeth McCown

Previous publications of the Society:

Perspectives on Human Evolution, Volume I.
S. L. Washburn and Phyllis C. Jay, Editors.
New York: Holt, Rinehart and Winston, 1968.

Perspectives on Human Evolution, Volume II.
S. L. Washburn and Phyllis Dolhinow, Editors.
New York: Holt, Rinehart and Winston, 1972.

Perspectives on Human Evolution, Volume III.
Glynn Ll. Isaac and Elizabeth R. McCown, Editors.
Menlo Park, California: W. A. Benjamin, Inc., 1975.

Preface

The purpose of the *Perspectives on Human Evolution* series is to bring together the variety of evidence which helps us to understand the process that has created mankind. Papers on the course of evolution, the fossil record, tools, and biosocial problems appear in each of the four volumes of the series. The second volume contains six papers on language alone, stressing the importance of this topic. The great significance of Africa forms the subject matter of Volume III.

In this fourth volume, the range of subjects includes natural selection, the brain, sexual behaviors, and the social system. The more general topics that have appeared in the previous volumes are supplemented by intriguing papers on the Piltdown fabrication (forging a head), and on toolmaking by orangutans. A very special contribution is the collection of drawings by Adolph Schultz, which illustrates many points in human evolution. The final chapter emphasizes the importance of biology for the future of social science.

The volume is committed to the development of a new biosocial science. It is hoped that *Perspectives IV* will be a step in the direction of a new synthesis, one that incorporates a very wide variety of subjects and techniques to help illuminate the problems of human behavior.

We particularly wish to thank Alice Davis for help in every phase of preparing and editing the volume and John Staples for major help in planning the publication.

We are indebted to all the authors for their cooperation and to those individuals and organizations who gave permission to reproduce papers and illustrative material.

E. R. McCown and S. L. Washburn

Contents

Introduction   1

Ernest Mayr:
    The Nature of the Darwinian Revolution   11

Paul D. MacLean:
    The Evolution of Three Mentalities   33

Ronald E. Myers:
    Comparative Neurology of Vocalization and Speech: Proof of a Dichotomy   59

Mary LeCron Foster:
    The Symbolic Structure of Primordial Language   77

Frank A. Beach:
    Human Sexuality and Evolution   123

Beatrix A. Hamburg:
    The Biosocial Bases of Sex Difference   155

R. V. S. Wright:
    Imitative Learning of a Flaked Stone Technology—the Case of an Orangutan   215

Wilton Marion Krogman:
    The Planned Planting of Piltdown: Who? Why?   239

Adolph H. Schultz:
    Illustrations of the Relation Between Primate Ontogeny and Phylogeny   255

S. L. Washburn and Elizabeth R. McCown:
  Human Evolution and Social Science   285

The Authors   297

Index   301

Introduction

Evolution is the process that created humankind. This view of life (Simpson 1964) is so different from the preceding theories that it has been accepted only by a small proportion of people. Even today there is much opposition to the teaching of Darwinism in schools, as Mayr explains in Chapter 1. The evolutionary point of view required a whole new world view, a rethinking of religion and philosophy. Humans were no longer the center of a small universe, but late arrivals on a tiny planet in a universe that has existed for billions of years and that extends millions of light years in space. Such a view of the universe is simply contrary to common sense and the experience of humankind. Before the advances of modern science, all peoples believed in gods, magic, and the traditions of their group. Knowledge was limited to what could be readily perceived, and it was interpreted in human ways. There were no mechanisms for measuring time and space or for understanding biological and chemical processes. Evolution is but one aspect of the scientific revolution, and, theoretically, it should be no more

Our particular thanks to Mrs. Alice Davis for her editorial assistance and to the Wenner-Gren Foundation for Anthropological Research, whose support made the research for this article possible.

disturbing than physics or astronomy, or any other branches of modern science that have changed human perceptions of the world. But, for many, evolution is upsetting precisely because it brings technical science to bear on human problems and on our place in nature.

People have to act, and social acts are embedded in tradition and based on emotion. It is natural to think of one's own beliefs as correct. And yet most human opinions have proven to be wrong—not wrong in some minor way, but deeply and profoundly in error. The prescientific world view was based on belief in miracles, magic, spontaneous generation, spirits, and the sanctity of local custom. The human brain, prior to experimental science, believed the grossest misconceptions with serene assurance, and philosophers still try to think their way to the solution of problems that are far beyond the capability of the unaided human brain (Rhinelander 1973). In Chapter 2 MacLean examines the brain from an evolutionary point of view and demonstrates its ability to invest thoughts with the feeling that they are "real, true, and important." This emotional cerebration, as MacLean calls it, is essential for survival. An evolutionary view suggests that the brain functions very differently from what our intellectual or common sense traditions might lead us to expect.

Evolution is difficult to accept, both because the evidence is technical and because it destroys a whole pattern of beliefs. An evolutionary view of the brain is hard for the academic community to accept because it challenges the beliefs upon which the intellectual life is founded. As MacLean (1970: 337) has put it, "emotional cerebration appears to have the paradoxical capacity to find equal support for the opposite sides of any question." Surely this ability is abundantly illustrated in the continuing controversies over human evolution! As one reads the scientific literature it is by no means clear where "science" ends and emotional cerebration begins (Washburn 1974).

Human communication—emotional, rational, or the usual mixture of the two—depends heavily on language. Although the multimodal communication—gesture plus sound—(Lancaster 1975) of nonhuman primates continues, human language adds new dimensions not seen in any other animal. Humans can communicate far more than any other primate, and clearly there has been selection for this adaptively important ability. Communication is made possible by a phonetic code, by short, contrasting, meaningless units combined into meaningful units. This system of communication has the characteristic of codes in general—a very small number of units may be combined in an almost infinite number of ways. It is the sound code that makes the human system of communication open, by making new and meaningful combinations of sounds (phonemes) possible. By contrast, the multimodal communication systems of nonhuman primates are closed—they lack the short contrasting units that may be built into new combinations. In Chapter 3 Myers discusses the anatomical basis for the uniquely human vocal-auditory system of communication. He shows that the basis for the human system is in the brain, and that comparative, experimental, and clinical information all indicate that the human vocal-auditory system is new and involves large parts of the brain. He also shows that humans continue to use the phylogenetically old multimodal system, primarily to communicate emotional states.

Myers's paper not only clarifies the differences between the communication of humans and of other primates, but also provides a model for the comparative method. The comparison of communications (sounds, gestures, facial expressions) requires information from a variety of sources, and the meaning of behavioral comparisons is never clear without experimental analysis. The neurological control of sounds and facial expressions in humans is very different from that of the nonhuman primates, and comparisons of behaviors that omit the fundamental biological bases are bound to be misleading. To be useful, the comparative study of behavior must consider biological mechanisms and must resort to experimental methods wherever these are possible.

Recently there has been renewed interest in the origins of language and language universals (Greenberg 1975). Solutions to both problems require an understanding of the brain. Apes, because they lack the necessary neural structures (Myers, Chapter 3), cannot be taught to speak (Kellogg 1968). Humans learn to speak so easily that, in fact, this kind of learning is almost impossible to prevent (Lenneberg 1967). The ease with which all populations learn to speak is a fundamental biological universal upon which other linguistic universals are based. Human universal cognitive abilities are the result of natural selection, and the presence of great linguistic skill in all contemporary peoples shows that communication by language has been of great adaptive importance in the history of *Homo sapiens*.

Foster (Chapter 4) shows that the languages of *Homo sapiens* are similar, not only because of the brain, the vocal-auditory channel, and some general structural features (Greenberg 1975), but because of common origin. Existing languages are so similar that it is possible to reconstruct both the phonetic structure and the meaning of primordial linguistic forms. Foster sees the origin of language as "undoubtedly the single most important advance in biological history. With language as symbol, other aspects of culture were free to develop concomitantly" (page 116). According to this view, the great increase in the rate of cultural change that started some forty or fifty thousand years ago is the result of the cognitive-linguistic abilities of *Homo sapiens*. There must have been earlier forms of linguistic communication that laid the evolutionary background for the language of *Homo sapiens*, but no traces of such forms have survived. There are no populations of humans with a radically different primitive language, nor are there populations that rely largely on some different form of communication, such as gesture. In spite of ingenious attempts (Lieberman 1975), it has not been possible to reconstruct the vocal tract of fossil forms (LeMay 1975). Baboons may make vowel-like sounds of the sort that might easily evolve into a phonetically useful code (Andrew 1975). The problem of the origin of language appears to be primarily a question of the evolution of the brain (Myers Chapter 3).

Beach (Chapter 5) points out that "the total behavioral repertoire of any species can be regarded as a mechanism, or aggregation of mechanisms, reflecting that species' successful solution of the problems of survival and reproduction" (page 138). The loss of female estrus and the evolution of culturally defined gender roles have been critical in the history of our species. Beach sees the evolution of the sexual division of labor as a biosocial problem in which physiological changes were linked in a feedback relation with successful social behaviors. The evolution of human sexuality cannot be understood apart from the whole problem of the evolution of human behavior, and in this the brain is critical. Whether we are considering scientific theories (Chapter 1), the triune brain (Chapter 2), language and the brain (Chapter 3), primordial language (Chapter 4), or adaptation through gender roles (Chapter 5), rational and emotional factors are mediated by the brain.

The relation of the brain to sex differences in behavior is further considered by Hamburg in Chapter 6. Hamburg discusses reproductive behavior, aggression, language, spatial behavior, and affectional systems, stressing that from both an evolutionary and a developmental point of view the issues are biosocial. Sex differences cannot be understood either by the study of anatomical-physiological factors or by behavioral investigations in isolation. The approach of sociobiology, or behavioral science, is necessary. Overall understanding is based on the contributions of many different techniques and on the combining of knowledge from many different departments. Whether one is interested in sociology, clinical psychiatry, or evolution, the basic problem is the same. An understanding of sex differences involves many factors. In her overview of these factors, Hamburg stresses that "the all-important brain matures in a biochemical and social environment which may be different for males and females" (page 198) and that "in every culture the rewards and costs have been different for males and females" (page 198). While focusing on evolution and the understanding that comes from the past, Hamburg emphasizes that the present is really new and that the purpose of behavioral science is to aid in the comprehension of the good life. Our complex biology exists in rapidly changing social situations, and the aim of sociobiology is to bring understanding of these realities into the service of mankind.

The theory of evolution through natural selection was a major revolution in human thinking; it changed the whole conception of humankind and of our place in the universe. The human brain, a product of the evolutionary process, is far from being a rational instrument or seat of the soul. Human beings adapted to the conditions of the world of hunters and gatherers through the use of knowledge, skills, and social life. In all of these areas the brain was critical, especially in the evolution of language. Human social life depends on language—both for the transmission of information and for the communication of emotions that are essential to any social system. Our uniquely human sex and gender roles are products of behavioral evolution, and biology may play a fundamental part in many customs, such as exogamy (Bischof 1975).

The importance of the brain, rather than the hand, in making stone tools has been demonstrated by an ingenious experiment devised by Wright. In Chapter 7 Wright shows that an orangutan is capable of making a stone tool, and using it, if it comprehends the idea. Although the orangutan hand is the least human of any of the contemporary apes (Schultz 1969), it is capable of grooming and precise manipulation of small objects (Tuttle 1969). Since the hand of *Australopithecus* was far more human in its proportions than that of an orangutan (Napier 1970), the progress in toolmaking over the last three million years is probably a result of the evolution of the brain and of intelligence. During this time the thumb became larger and more powerful, and the fingers somewhat shorter (Schultz, Figure 5), but the main changes were in the areas of the brain controlling hand skills. Wright's evidence clearly shows that the only way to determine what apes can or cannot do is by experiment. It is sobering to remember that anthropologists once thought that no creature with a brain as small as *Australopithecus* was capable of making a tool. Brain size relative to body size is shown in Figure 1, from Jerison (1973).

The contrast between today's methods and the traditional study of human evolution could not be better illustrated than by the Piltdown problem. In Chapter 8 Krogman reviews the history of this celebrated forgery. In spite of the supposedly great importance of the discovery, no major excavations were ever made; the Piltdown skull and jaw were almost a surface find. However, over a 40-year period, comparative anatomical studies could not *prove* whether the jaw and skull belonged to one individual, or whether Piltdown was a "chimera," as Weidenreich maintained. In retrospect, it is the motivations of the participants which fascinate. Whatever the intent of the forger may have been, there was a great desire to believe in the authenticity of the "Earliest Englishman" (Smith Woodward 1948).

Although new techniques, a new generation of scientists, and a very different climate of opinion all combined to show that Piltdown was not a genuine fossil, Keith's exposition of how he reconstructed the pieces re-

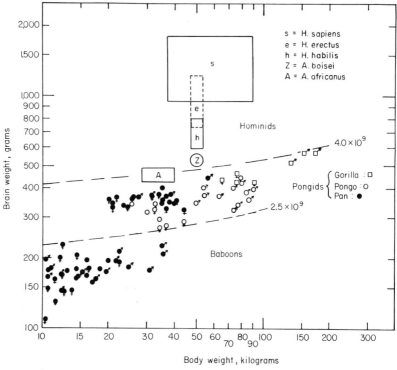

FIGURE 1

Brain: body proportions of baboons, apes, and hominids. The data on hominids have been expanded from the original version by Dr. Jerison. The brain size of the early hominids (*Australopithecus* in the widest sense) is a little larger than the apes and much smaller than later hominids.

mains most instructive. (Keith 1932). Keith's detailing of the reasons for each stage in the reconstruction provides invaluable insights into the nature of traditional anatomical reasoning. In retrospect it can be seen that, despite Keith's knowledge and reasoning, the parts were so broken that reconstruction was practically impossible. As Miller wrote in 1915, "Deliberate malice could hardly have been more successful than the hazards of deposition in so breaking the fossils as to give free scope to individual judgment in fitting the parts together" (p. 1). And yet the faith in anatomical reconstruction was so great that, as Krogman noted (page 242), the palate was described as more modern than the mandible. Of course, no palate was ever found.

The thirty illustrations of the "Schultz Atlas" represent many of Adolph

Schultz's best delineations of primate anatomy. They also show his major research interests in ontogenetic development, limb proportions, skull proportions, and dental variations. The original publications in which these illustrations appeared range over a period of some 50 years. No one has contributed more to the understanding of the primates, and this atlas is a monument to Schultz's unwavering interest and his great artistic skill.

Over the last few years there has been a great increase of interest in animal behavior. This first expressed itself in the proliferation of television programs, popular books, and the expanding investigations of the behaviors of many kinds of animals. Claims that such studies might be important in understanding human behavior received little attention until the publication of Wilson's *Sociobiology* (1975), a book that dramatized the possibility of applying modern genetic theory (inclusive fitness theory) to the interpretation of the behavior of all animals, including humans. The boldness and simplicity of the theory gave sociobiology its great appeal. Objections have arisen primarily because of the difficulty of applying the general theory to particular situations. In the interpretation of human behavior strong criticisms have been raised against postulating genes, minimizing the differences between learning and inheritance, and guessing at the function of customs.

No one doubts the importance of theory in influencing the directions of research, and much of sociobiology is a continuation of long-standing interest in evolution, behavior, and selection, put in modern and dramatic form. Without belittling genetic theory in any way, we think that behavioral science must be built on the understanding of the way the body functions. If the human brain is critically important in the analysis of human behaviors, then any theory that minimizes biological understanding of the brain will not be very useful. This is the assumption lying behind the choice of papers on the brain, language, and hand skills.

The last chapter is based on the belief that an analysis of the biology of the actors is essential to the full understanding of any social system. This is not contrary to the tradition, stemming from Durkheim, that social facts form a logical universe which can only be explained by social facts. The assumption is simply that a particular species *(Homo sapiens)* is necessary before there can be any social facts and that biology inevitably enters into the foundations of social facts. From a practical point of view this is seen most clearly in medicine. But behind the vast diversity of human social systems are the individual organisms who think, create, speak, and feel. In the long run, evolution lies behind the social facts, and in the short run individual organisms carry and modify the behaviors the scientist labels as social facts.

We think that the social sciences have been separated long enough and that it is now useful to consider the nature of the biology of the human actors in the social systems. Clearly this will only be useful for some specific problems and will only supplement the tradition of Durkheim, not supplant it. The world of many cultures is rapidly disappearing, and the determination of what will constitute reasonable social facts in the future cannot be determined without considering the nature of human beings.

References Cited

Andrew, R. J.
1976   Use of formants in the grunts of baboons and other nonhuman primates. *In* Origins and Evolution of Language and Speech. Annals of the New York Academy of Sciences, Vol. 280, pp. 673–693.

Bischof, N.
1975   Comparative ethology of incest avoidance. *In* Biosocial Anthropology. R. Fox, ed. Pp. 37–67. London: Malaby Press.

Fox, R., ed.
1975   Biosocial Anthropology. London: Malaby Press.

Greenberg, J. H.
1975   Research on language universals. In Annual Review of Anthropology. B. J. Siegel, A. R. Beals, and S. A. Tyler, eds. Palo Alto, California: Annual Reviews.

Jerison, H. J.
1973   Evolution of the Brain and Intelligence. New York: Academic Press.

Keith, A.
1932   The Antiquity of Man, Vol. 1. Philadelphia: Lippincott.

Kellogg, W. N.
1968   Communication and language in the home-raised chimpanzee. Science 162:423–427.

LeMay, Marjorie
1975   The language capabilities of Neanderthal Man. American Journal of Physical Anthropology 42:9–14.

Lenneberg, E. H.
1967   Biological Foundations of Language. New York: John Wiley.

MacLean, Paul D.
1970   The triune brain, emotion and scientific bias. *In The Neurosciences: Second Study Program.* F. O. Schmitt, ed. Pp. 336–349. New York: Rockefeller University Press.

Miller, G. S.
1915   The Jaw of the Piltdown Man. Smithsonian Miscellaneous Collections, Vol. 63, pp. 1–31.

Napier, J. R.
1970   The Roots of Mankind. Washington, D.C.: Smithsonian Institution.

Oakley, K. P.
    1972   Skill as a human possession. *In* Perspectives on Human Evolution. S. L. Washburn and P. Dolhinow, eds. Pp. 14–50. New York: Holt, Rinehart and Winston.

Rhinelander, P. H.
    1973   Is Man Incomprehensible to Man? San Francisco: W. H. Freeman.

Schultz, A. H.
    1969   The Life of Primates. London: Weidenfeld and Nicolson.

Simpson, G. G.
    1964   This View of Life. New York: Harcourt, Brace and World.

Smith Woodward, A.
    1948   The Earliest Englishman. London: Watts and Co. (The Thinker's Library, No. 127.)

Tuttle, R. H.
    1969   Quantitative and functional studies on the hands of the Anthropoidea. Journal of Morphology 128:309–364.

Washburn, S. L.
    1974   Human evolution: science or game? *In* Yearbook of Physical Anthropology 1973. Yearbook Series 17:67–70.

Washburn, S. L., and E. R. McCown
    1972   Evolution of human behavior. Social Biology 19:163–170.

Wilson, E. O.
    1975   Sociobiology. Cambridge, Massachusetts: Harvard University and Belknap Press.

Relative importance of various vertebrates since the Cambrian period. (Redrawn from *Biology: Its Human Implications* by Garrett Hardin. W. H. Freeman and Company. Copyright © 1949.)

Date in millions of years	Group of animals
480	Sharks
420	Bony fish
400	Amphibians
300	Reptiles
250	Birds
200	
170	Mammals
125	Man
65	

Ernst Mayr

The Nature of the Darwinian Revolution

Many groups are still opposed to the theory of evolution. Religious opposition has been regarded as the main reason for the slow acceptance of Darwin's ideas, but Mayr shows that a number of concepts and beliefs made acceptance of the Darwinian revolution difficult. The revolution was far more complicated than the replacement of a single notion, and for a long time even scientists fought the idea and consequences of natural selection.

The road on which science advances is not a smoothly rising ramp; there are periods of stagnation, and periods of accelerated progress. Some historians of science have recently emphasized that there are occasional breakthroughs, scientific revolutions(1), consisting of rather drastic revisions of previously maintained assumptions and concepts. The actual nature of these revolutions, however, has remained highly controversial(2). When we look

This article has been reprinted, with permission, from *Science,* Volume 176, pages 981–989. Italicized numbers in parentheses refer to the sources and notes listed at the end of the article.

at those of the so-called scientific revolutions that are most frequently mentioned, we find that they are identified with the names Copernicus, Newton, Lavoisier, Darwin, Planck, Einstein, and Heisenberg; in other words, with one exception, all of them are revolutions in the physical sciences.

Does this focus on the physical sciences affect the interpretation of the concept "scientific revolution"? I am taking a new look at the Darwinian revolution of 1859, perhaps the most fundamental of all intellectual revolutions in the history of mankind. It not only eliminated man's anthropocentrism, but affected every metaphysical and ethical concept, if consistently applied. The earlier prevailing concept of a created, and subsequently static, world was miles apart from Darwin's picture of a steadily evolving world. Kuhn (1) maintains that scientific revolutions are characterized by the replacement of an outworn paradigm by a new one. But a paradigm is, so to speak, a bundle of separate concepts, and not all of these are changed at the same time. In this analysis of the Darwinian revolution, I am attempting to dissect the total change of thinking involved in the Darwinian revolution into the major changing concepts, to determine the relative chronology of these changes, and to test the resistance to these changes among Darwin's contemporaries.

The idea of evolution had been widespread for more than 100 years before 1859. Evolutionary interpretations were advanced increasingly often in the second half of the 18th and the first half of the 19th centuries, only to be ignored, ridiculed, or maligned. What were the reasons for this determined resistance?

The history of evolutionism has long been a favorite subject among historians of science (3–5). Their main emphasis, however, has been on Darwin's forerunners, and on any and every trace of evolutionary thinking prior to 1859, or on the emergence of evolutionary concepts in Darwin's own thinking. These are legitimate approaches, but it seems to me that nothing brings out better the revolutionary nature of some of Darwin's concepts (6) than does an analysis of the arguments of contemporary antievolutionists.

Cuvier, Lyell, and Louis Agassiz, the leading opponents of organic evolution, were fully aware of many facts favoring an evolutionary interpretation, and likewise of the Lamarckian and other theories of transmutation. They devoted a great deal of energy to refute evolutionism (7–10) and supported instead what, to a modern student, would seem a less defensible position. What induced them to do so?

It is sometimes stated that they had no legitimate choice, because—it is claimed—not enough evidence in favor of evolution was available before 1859. The facts refute this assertion. Lovejoy (11), in a superb analysis of this question, asks: "At what date can the evidence in favor of the theory of organic evolution . . . be said to have been fairly complete?" Here, one can perhaps distinguish two periods. During an earlier one, lasting from about 1745 to 1830, much became known that suggested evolution or, at least, a temporalized scale of perfection *(12)*. Names like Maupertuis (1745), de Maillet (1749), Buffon (1749), Diderot (1769), Erasmus Darwin (1794), Lamarck (1809), and E. Geoffrey St. Hilaire (1818) characterize this period. Enough evidence from the fields of biogeography, systematics, paleontology, comparative anatomy, and animal and plant breeding, was already available by about 1812 (date of Cuvier's *Ossemens Fossiles*) to have made it possible to develop some of the arguments later made by Darwin in the *Origin of Species (6)*. Soon afterward, however, much new evidence was produced by paleontology and stratigraphy, as well as by biogeography and comparative anatomy, with which only the evolutionary hypothesis was consistent; these new facts "reduced the rival hypothesis to a grotesque absurdity" *(11)*. Yet, only a handful of authors [including Meckel (1821), Chambers (1844), Unger (1852), Schaaffhausen (1853), Wallace (1855)] adopted the concept of evolution while such leading authorities as Lyell, R. Owen, and Louis Agassiz vehemently opposed it.

Time does not permit me to marshal the abundant evidence in favor of evolution which existed by 1830. A comprehensive listing has been provided by Lovejoy *(11)*, although the findings of systematics and biogeography must be added to his tabulation. The patterns of animal distribution were particularly decisive evidence, and it is no coincidence that Darwin devoted to it two entire chapters in the *Origin*. In spite of this massive evidence, creationism remained "the hypothesis tenaciously held by most men of science for at least twenty years before 1859" *(11)*. It was not a lack of supporting facts, then, that prevented the acceptance of the theory of evolution, but rather the power of the opposing ideas.

Curiously, a number of nonscientists, particularly Robert Chambers *(13)* and Herbert Spencer, saw the light well before the professionals. Chambers, the author of the *Vestiges of the Natural History of Creation,* developed quite a consistent and logical argument for evolutionism, and was instrumental in converting A. R. Wallace, R. W. Emerson, and A. Schopenhauer to evolutionism. As was the case with Diderot and Erasmus Darwin, these well-informed and broadly educated lay people looked at the problem in a "holistic" way, and thus perceived the truth more readily than did the professionals who were committed to certain well-established dogmas. A view from the distance is sometimes more revealing, for the understanding of broad issues, than the myopic scrutiny of the specialist.

POWER OF RETARDING CONCEPTS

Why were the professional geologists and biologists so blind when the manifestations of evolution were staring them in the face from all directions? Darwin's friend Hewett Watson put it this way in 1860 *(14,* p. 226):

"How could Sir Lyell . . . for thirty years read, write, and think on the subject of species *and their succession,* and yet constantly look down the wrong road?" Indeed, how could he? And the same question can be asked for Louis Agassiz, Richard Owen, almost all of Lyell's geological colleagues, and all of Darwin's botanist friends from Joseph Hooker on down. They all displayed a nearly complete resistance to drawing what to us would seem to be the inevitable conclusion from the vast amount of evidence in favor of evolution.

Historians of science are familiar with this phenomenon; it happens almost invariably when new facts cast doubt on a generally accepted theory. The prevailing concepts, although more difficult to defend, have such a powerful hold over the thinking of all investigators that they find it difficult, if not impossible, to free themselves of these ideas. To illustrate this by merely one example, I would like to quote a statement by Lyell: "It is idle . . . to dispute about the abstract possibility of the conversion of one species into another, when there are known causes, so much more active in their nature, which must always intervene and prevent the actual accomplishment of such conversions" (9, p. 162). Actually, one searches in vain for a demonstration of such "known causes" and any proof that they "must" always intervene. The cogency of the argument relied entirely on the validity of silent assumptions.

In the particular case of the Darwinian revolution, what were the dominant ideas that formed roadblocks against the advance of evolutionary thinking? To name these concepts is by no means easy because they are silent assumptions, never fully articulated. When these assumptions rest on religious beliefs or on the acceptance of certain philosophies, they are particularly difficult to reconstruct. This is the major reason why there is so much difference of opinion in the interpretation of this period. Was theology responsible for the lag, or was it the authority of Cuvier or Lyell, or the acceptance of catastrophism (with progressionism), or the absence of a reasonable explanatory scheme? All of these interpretations and several others have been advanced, and all presumably played some role. Others, particularly the role of essentialism, have so far been rather neglected by the historians.

NATURAL THEOLOGY AND CREATIONISM

The period from 1800 to the middle of that century witnessed the greatest flowering of natural theology in Great Britain (5, 15). It was the age of Paley and the Bridgewater Treatises, and virtually all British scientists accepted the traditional Christian conception of a Creator God. The industrial revolution was in full swing, the poor workingman was exploited unmercifully, and the goodness and wisdom of the Creator was emphasized constantly to sooth guilty consciences. It became a moral obligation for the scientist to find additional proofs for the wisdom and constant attention of the Creator. When Chambers in his *Vestiges (13)* dared to replace direct intervention of the Creator by the action of secondary causes (natural laws), he was roundly condemned. Although the attacks were ostensibly directed against errors of fact, virtually all reviewers were horrified that Chambers had "annulled all distinction between physical and moral," and that he had degraded man by ranking him as a descendant of the apes and by interpreting the universe as "the progression and development of a rank, unbending, and degrading materialism" (*5*, p. 150; *16*). It is not surprising that in this intellectual climate Chambers had taken the precaution of publishing anonymously. Yet the modern reader finds little that is objectionable in Chambers's endeavor to replace supernatural explanations by scientific ones.

To a greater or lesser extent, all the scientists of that period resorted, in their explanatory schemes, to frequent interventions by the Creator (in the running of His world). Indeed, proofs of such interventions were considered the foremost evidence for His existence. Agassiz quite frankly describes the obligations of the naturalist in these words: "Our task is . . . complete as soon as we have proved His existence" (*10*, p. 132). To him the *Essay on Classification* was nothing but another Bridgewater Treatise in which the relationship of animals supplied a particularly elaborate and, for Agassiz, irrefutable demonstration of His existence.

Natural theology equally pervades Lyell's *Principles of Geology*. After discussing various remarkable instincts, such as pointing and retrieving, which are found in races of the dog, Lyell states: "When such remarkable habits appear in races of this species, we may reasonably conjecture that they were given with no other view than for the use of man and the preservation of the dog which thus obtains protection" (*9*, p. 455). Even though cultivated plants and domestic animals may have been created long before man, "some of the qualities of particular animals and plants may have been given solely with a view to the connection which, it was foreseen, would exist between them and man" (*9*, p. 456). Like Agassiz, Lyell believed that everything in nature is planned, designed, and has a predetermined end. "The St. Helena plants and insects [which are now dying out] may have lasted for their allotted term" (*9*, p. 9). The harmony of living nature and all the marvelous adaptations of animals and plants to each other and to their environment seemed to him thus fully and satisfactorily explained.

CREATIONISM AND THE ADVANCES OF GEOLOGICAL SCIENCE

At the beginning of the 18th century, the concept of a created world seemed internally consistent as long as this world was considered only recently created (in 4004 B.C.), static, and unchanging. The "ladder of perfection" (part of God's plan) accounted for the "higher" and "lower" organization of animals and man, and Noah's flood for the existence of fossils. All this could be readily accommodated within the framework of a literal biblical interpretation.

The discovery of the great age of the earth *(5, 17)* and of an ever-increasing number of distinct fossil faunas in different geological strata necessitated abandoning the idea of a single creation. Repeated creations had to be postulated, and the necessary number of such interventions had to be constantly revised upward. Agassiz was willing to accept 50 or 80 total extinctions of life and an equal number of new creations. Paradoxically, the advance of scientific knowledge necessitated an increasing recourse to the supernatural for explanation. Even such a sober and cautious person as Charles Lyell frequently explained natural phenomena as due to "creation" and, of course, a carefully thought-out creation. The fact that the brain of the human embryo successively passes through stages resembling the brains of fish, reptile, and lower mammal discloses, "in a highly interesting manner, the unity of plan that runs through the organization of the whole series of vertebrated animals; but [it] lend[s] no support whatever to the notion of a gradual transmutation of one species into another; least of all of the passage, in the course of many generations, from an animal of a more simple to one of a more complex structure" *(9,* p. 20). When a species becomes extinct it is replaced "by new creations" *(9,* p. 45). Nothing is impossible in creation. "Creation seems to require omnipotence, therefore we cannot estimate it" *(18,* p. 4). "Each species may have had its origin in a single pair, or individual where an individual was sufficient, and species may have been created in succession at such times and in such places as to enable them to multiply and endure for an *appointed* period, and occupy an *appointed* space of the globe." (italics mine) *(9,* pp. 99–100). Everything is done according to plan. Since species are fixed and unchangeable, everything about them, such as the area of distribution, the ecological context, adaptations to cope with competitors and enemies, and even the date of extinction, was previously "appointed," that is, predetermined.

This constant appeal to the supernatural amounted to a denial of all sound scientific methods, and to the adoption of explanations that could neither be proven nor refuted. Chambers saw this quite clearly *(13).* When there is a choice between two theories, either special creation or the operation of general laws instituted by the Creator, he exclaimed, "I would say that the latter [theory] is greatly preferable, as it implies a far grander view of the Divine power and dignity than the other" *(13,* p. 117). Indeed, the increasing knowledge of geological sequences, and of the facts of comparative anatomy and geographic distribution, made the picture of special creation more ludicrous every day *(11,* p. 413).

ESSENTIALISM AND A STATIC WORLD

Thus, theological considerations clearly played a large role in the resistance to the adoption of evolutionary views in England (and also in France). Equally influential, or perhaps even more so, was a philosophical concept. Philosophy and natural history during the first half of the 19th century, particularly in continental Europe, were strongly dominated by typological thinking [designated "essentialism" by Popper *(19, 20)*]. This presumes that the changeable world of appearances is based on underlying immutable essences, and that all members of a class represent the same essence. This idea was first clearly enunciated in Plato's concept of the *eidos*. Later it became a dominant element in the teachings of Thomism *(21),* and of all idealistic philosophy. The enormous role of essentialism in retarding the acceptance of evolutionism was long overlooked*(22, 23)*. The observed vast variability of the world has no more reality, according to this philosophy, than the shadows of an object on a cave wall, as Plato expressed it in his allegory. The only things that are permanent, real, and sharply discontinuous from each other are the fixed, unchangeable "ideas" underlying the observed variability. Discontinuity and fixity are, according to the essentialist, as much the properties of the living as of the inanimate world.

As Reiser*(24)* has said, a belief in discontinuous, immutable essences is incompatible with a belief in evolution. Agassiz was an extreme representative of this philosophy *(23)*. To a lesser extent the same can be demonstrated for all of the other opponents of evolutionism, including Lyell. When rejecting Lamarck's claim that species and genera intergrade with each other, Lyell proposes that the following laws "prevail in the economy of the animate creation. . . . Thirdly, that there are fixed limits beyond which the descendants from common parents can never deviate from a certain type; fourthly, that each species springs from one original stock, and can never be permanently confounded by intermixing with the progeny of any other stock; fifthly, that each species shall endure for a considerable period of time" (9, p. 433). All nature consists, according to Lyell, of fixed types created at a definite time. To him these types were morphological entities, and he was rather shocked by Lamarck's idea that changes in behavior could have any effect on morphology.

As an essentialist, Lyell showed no understanding of the nature of genetic variation. Strictly in the scholastic tradition, he believed implicitly that essential characters could not change; this could occur only with non-essential characters. If an animal is brought into a new environment, "a short period of time is generally sufficient to effect nearly the whole change which an alteration of external circumstances can bring about in the habits of a species. . . . such capacity of accommodation to new circumstances is enjoyed in very different degrees by different species" (9, p. 464). For instance, if we look at the races of dogs, they show many superficial differences "but, if we look for some of those essential changes which would be required to lend even the semblance of a foundation for the theory of Lamarck, respecting the growth of new organs and the gradual obliteration of others, we find nothing of the kind" (9, p. 438). This forces Lyell to question even Lamarck's conjecture "that the wolf may have been the original of the dog." The fact that in the (geologically speaking) incredibly short time since the dog was domesticated, such drastically different races as the Eskimo dog, the hairless Chihuahua, the greyhound, and other extremes have evolved is glossed over.

LYELL'S SPECIES CONCEPT

Holding a species concept that allowed for no essential variation, Lyell credited species with little plasticity and adaptability. This led him to an interpretation of the fossil record that is very different from that of Lamarck. Anyone studying the continuous changes in the earth's surface, states Lyell, "will immediately perceive that, amidst the vicissitudes of the earth's surface, species cannot be immortal, but must perish, one after the other, like the individuals which compose them. There is no possibility of escaping from this conclusion, without resorting to some hypothesis as violent as that of Lamarck who imagined . . . that species are each of them endowed with indefinite powers of modifying their organization, in conformity to the endless changes of circumstances to which they are exposed" (9, pp. 155–156).

The concept of a steady extermination of species and their replacement by newly created ones, as proposed by Lyell, comes close to being a kind of microcatastrophism, as far as organic nature is concerned. Lyell differed from Cuvier merely in pulverizing the catastrophes into events relating to single species, rather than to entire faunas. In the truly decisive point, the rejection of any possible continuity between species in progressive time sequences, Lyell entirely agreed with Cuvier, When he traced the history of a species backward, Lyell inexorably arrived at an original ancestral pair, at the original center of creation. There is a total absence in his arguments of any thinking in terms of populations.

The enormous power of essentialism is in part explainable by the fact that it fitted the tenets of creationism so well; the two dogmas strongly reinforced each other. Nothing in Lyell's geological experience seriously contradicted his essentialism. It was not shaken until nearly 25 years later when Lyell visited the Canary Islands (from December 1853 to March 1854) and became acquainted with the same kind of phenomena that, in the Galapagos, had made Darwin an evolutionist and which, in the East Indian

Archipelago, gave concrete form to the incipient evolutionism of A. R. Wallace. Wilson *(18)* has portrayed the growth of doubt which led Lyell to publicly confess his conversion to evolutionism in 1862. The adoption of population thinking by him was a slow process, and even years after his memorable discussion with Darwin (16 April 1856), Lyell spoke in his notebooks of "variation *or* selection" as the important factor in evolution in spite of the fact that Darwin's entire argument was founded on the need for *both* factors as the basis of a satisfactory theory.

LYELL AND UNIFORMITARIANISM

It is a long-standing tradition in biological historiography that Lyell's revival of Hutton's theory of uniformitarianism was a major factor in the eventual adoption of evolutionary thinking. This thesis seems to be a great oversimplification; it is worthwhile to look at the argument a little more critically *(25)*. When the discovery of a series of different fossil faunas, separated by unconformities, made the story of a single flood totally inadequate, Cuvier and others drew the completely correct conclusion that these faunas, particularly the alternation of marine and terrestrial faunas, demonstrated a frequent alternation of rises of the sea above the land and the subsequent reemergence of land above the sea. The discovery of mammoths frozen into the ice of Siberia favored the additional thesis that such changes could happen very rapidly. Cuvier was exceedingly cautious in his formulation of the nature of these "revolutions" and "catastrophes," but he did admit, "The breaking to pieces and overturning of the strata, which happened in former catastrophes, show plainly enough that they were sudden and violent like the last [which killed the mammoths and embedded them in ice]" (*26*, p. 16). He implied that most of these events were local rather than universal phenomena, and he did not maintain that a new creation had been required to produce the species existing today. He said merely "that they [modern species] did not anciently occupy their present locations and that they must have come there from elsewhere (*26*, pp. 125–126).

Cuvier's successors did not maintain his caution. The school of the so-called progressionists *(27)* postulated that each fauna was totally exterminated by a catastrophe at the end of each geologic period, followed by the special creation of an entirely new organic world. Progressionism, therefore, was intellectually a backward step from the widespread 18th century belief that the running of the universe required only occasional, but definitely not incessant, active intervention by the Creator: He maintained stability largely through the laws that He had decreed at the beginning, and which allowed for certain planetary and other perturbations. This same reasoning could have easily been applied to the organic world, and this indeed is what was done by Chambers in 1844, and by many other devout Christians after 1859.

Catastrophism was not as great an obstacle to evolutionism as often claimed. It admitted, indeed it emphasized, the advance which each new creation showed over the preceding one. By also conceding that there had been 30, 50, or even more than 100 extinctions and new creations, it made the concept of these destructions increasingly absurd, and what was finally left, after the absurd destructions had been abandoned, was the story of the constant progression of faunas *(28)*. As soon as one rejected reliance on supernatural forces, this progression automatically became evidence in favor of evolution. The only other assumption one had to make was that many of the catastrophes and extinctions had been localized events. This was, perhaps, not too far from Cuvier's original viewpoint.

The reason why catastrophism was adopted by virtually all of the truly productive leading geologists in the first half of the 19th century is that the facts seemed to support it. Breaks in fossil strata, the occurrence of vast lava flows, a replacement of terrestrial deposits by marine ones and the reverse, and many other phenomena of a similar, reasonably violent nature (including the turning upside down of whole fossil sequences) all rather decisively refuted a rigid uniformitarian interpretation. This is why Cuvier, Sedgwick, Buckland, Murchison, Conybeare, Agassiz, and de Beaumont, to mention a few prominent geologists, adopted more or less catastrophist interpretations.

Charles Lyell was the implacable foe of the "catastrophists," as his opponents were designated by Whewell *(29)*. In his *Principles of Geology (9)*, Lyell promoted a "steady state" concept of the world, best characterized by Hutton's motto, "no vestige of a beginning—no prospect of an end." Whewell coined the term "uniformitarianism" *(30)* for this school of thought, a term which unfortunately had many different meanings. The most important meaning was that it postulated that no forces had been active in the past history of the earth that are not also working today. Yet, even this would permit two rather different interpretations. Even if one includes supernatural agencies among forces and causes, one can still be a consistent uniformitarian, provided one postulates that the Creator continues to reshape the world actively even at the present. Rather candidly, Lyell refers to this interpretation, accepted by him, as "the perpetual intervention hypothesis" *(18,* p. 89).

Almost diametrically opposed to this were the conclusions of those who excluded all recourse to supernatural interventions. Uniformitarianism to them meant simply the consistent application of natural laws not only to inanimate nature (as was done by Lyell) but also to the living world (as proposed by Chambers). The important component in their argument was the rejection of supernatural intervention rather than a lip service to the word uniformity.

It is important to remember that Lyell applied his uniformitarianism in a consistent manner only to inanimate nature, but left the door open for special creation in the living world. Indeed, as Lovejoy (11) states justly, when it came to the origin of new species, Lyell, the great champion of uniformitarianism, embraced "the one doctrine with which uniformitarianism was wholly incompatible—the theory of numerous and discontinuous miraculous special creations." Lyell himself did not see it that way. As he wrote to Herschel (31), he considered his notion "of a succession of extinction of species, and creation of new ones, going on perpetually now . . . the grandest which I had ever conceived, so far as regards the attributes of the Presiding Mind." There is evidence, however, that Lyell considered these creations not always as miracles, but sometimes as occurring "through the intervention of intermediate causes" thus being "a natural, in contradistinction to a miraculous process." By July 1856, after having read Wallace's 1855 paper, and after having discussed evolution with Darwin (16 April 1856), Lyell had become completely converted to believing that the introduction of new species was "governed by laws in the same sense as the Universe is governed by laws" (18, p. 123).

Only the steady-state concept of uniformitarianism was novel in Lyell's interpretation. The insistence that nature operates according to eternal laws, with the same forces acting at all times was, from Aristotle on, the standard explanation among most of those who did not postulate a totally static world, for instance, among the French naturalists preceding Cuvier. Consequently, acceptance of uniformitarianism did not, as Lyell himself clearly demonstrated, require the acceptance of evolutionism. If one believed in a steady-state world, as did Lyell, uniformitarianism was incompatible with evolution. Only if it was combined with the concept of a steadily changing world, as it was in Lamarck's thinking, did it encourage a belief in evolution. It is obvious, then, that the statement "uniformitarianism is the pacemaker of evolutionism," is an exaggeration, if not a myth.

But what effect did Lyell have on Darwin? Everyone agrees that it was profound; there was no other person whom Darwin admired as greatly as Lyell. *Principles of Geology,* by Lyell, was Darwin's favorite reading on the *Beagle* and gave his geological interests new direction. After the return of the *Beagle* to England, Darwin received more stimulation and encouragement from Lyell than from any other of his friends. Indeed, Lyell became a father figure for him and stayed so for the rest of his life. Darwin's whole way of writing, particularly in the *Origin of Species,* was modeled after the *Principles.* There is no dispute over these facts.

But, what was Lyell's impact on Darwin's evolutionary ideas? There is much to indicate that the influence was largely negative. Knowing how firmly Lyell was opposed to the possibility of a transmutation of species, as documented by his devastating critique of Lamarck, Darwin was very careful in what he revealed to Lyell. He admitted that he doubted the fixity of species, but after that the two friends apparently avoided a further discussion of the subject. Darwin was far more outspoken with Hooker, to whom he confessed as early as January 1844, "I am almost convinced . . . that species are not (it is like confessing murder) immutable" (*14*, p. 23). It was not until 1856 that Darwin fully outlined his theory of evolution to Lyell (*18*, p. xlix). This reticence of Darwin was not due to any intolerance on Lyell's part (or else Lyell would not have, after 1856, encouraged Darwin so actively to publish his heretical views), but rather to an unconscious fear on Darwin's part that his case was not sufficiently persuasive to convert such a formidable opponent as Lyell. There has been much speculation as to why Darwin had been so tardy about publishing his evolutionary views. Several factors were involved (one being the reception of the *Vestiges*), but I am rather convinced that his awe of Lyell's opposition to the transmutation of species was a much more weighty reason than has been hitherto admitted. It is no coincidence that Darwin finally began to write his great work within three months after Lyell took the initiative to consult him and to encourage him. Lovejoy summarizes the effect of Lyell's opposition to evolution in these words: "It was . . . his example and influence, more than the logical force of his arguments, that so long helped to sustain the prevalent belief that transformism was not a scientifically respectable theory" (*11*). I entirely agree with this evaluation.

UNSUCCESSFUL REFUTATIONS OWING TO WRONG CHOICE OF ALTERNATIVES

Creationism, essentialism, and Lyell's authority were not, however, the only reasons for the delay in the acceptance of evolution; others were important weaknesses in the scientific methodology of the period. There was still a demand for conclusive proofs. "Show me the breed of dogs with an entirely new organ," Lyell seems to say, "and I will believe in evolution." That much of science consists merely in showing that one interpretation is more probable than another one, or consistent with more facts than another one, was far less realized at that period than it is now (*32*).

That victory over one's opponent consists in the refutation of his arguments, however, was taken for granted. Cuvier's, Lyell's, Agassiz's, and Darwin's detailed argumentations were all attempts to "falsify," as Popper (*33*) has called it, the statements of their opponents. This method, however, has a number of weaknesses. For instance, it is often quite uncertain what kind of evidence or argument truly represents a falsification. More fatal is the frequently made assumption that there are only two alternatives in a dispute. Indeed, the whole concept of "alternative" is rather ambiguous, as I shall try to illustrate with some examples from pre-Darwinian controversies.

We can find numerous illustrations in the antievolutionary writings of Charles Lyell and Louis Agassiz of the limitation to only two alternatives when actually there was at least a third possible choice. Louis Agassiz, for instance, never seriously considered the possibility of true evolution, that is, of descent with modification. For him the world was either planned by the Creator, or was the accidental product of blind physical causes (in which case evolution would be the concatenation of such accidents). He reiterates this singularly simple-minded choice throughout the *Essay on Classification* *(10):* "physical laws" versus "plan of creation" (p. 10), "spontaneous generation" versus "divine plan" (p. 36), "physical agents" versus "plan ordained from the beginning" (p. 37), "physical causes" versus "supreme intellect" (p. 64), and "physical causes" versus "reflective mind" (p. 127). By this choice he not only excluded the possibility of evolution as envisioned by Darwin, but even as postulated by Lamarck. Nowhere does Agassiz attempt to refute Lamarckian evolution. His physical causes, in turn, are an exceedingly narrow definition of natural causes, since it is fully apparent that Agassiz had a very simple-minded Cartesian conception of physical causes as motions and mechanical forces. "I am at a loss to conceive how the origin of parasites can be ascribed to physical causes" *(10,* p. 126). "How can physical causes be responsible for the form of animals when so many totally different animal types live in the same area subjected to identical physical causes?" *(10,* pp. 13–14). The abundant regularities in nature demonstrate "the plan of a Divine Intelligence" since they cannot be the result of blind physical forces. (This indeed was a standard argument among adherents of natural theology.) It never occurred to Agassiz that none of his arguments excluded a third possibility, the gradual evolution of these regularities by processes that can be daily observed in nature. This is why the publication of Darwin's *Origin* was such a shock to him. The entire evidence against evolution, which Agassiz had marshaled so assiduously in his *Essay on Classification,* had become irrelevant. He had failed completely to provide arguments against a third possibility, the one advanced by Darwin.

The concept of evolution, at that period, still evoked in most naturalists the image of the *scala naturae,* the ladder of perfection. No one was more opposed to this concept than Lyell, the champion of a steady-state world. Any finding that contradicted a steady progression from the simple toward the more perfect refuted the validity of evolution, he thought. Indeed, the fact that mammals appeared in the fossil record before birds, and that primates appeared in the Eocene considerably earlier than some of the orders of "lower" mammals were, to him, as decisive a refutation of the evoluionary theory as was to Agassiz the fact that the four great types of animals appeared simultaneously in the earliest fossil-bearing strata.

The assumption that refuting the *scala naturae* would refute once and for all any evolutionary theory is another illustration of insufficient alternatives. Lyell was quite convinced that the concept of a steady-state world would be validated (including regular special creations), if it could be shown that those mechanisms were improbable or impossible which Lamarck had proposed to account for the evolutionary change.

But there were also other violations of sound scientific method; for instance, the failure to see that both of two alternatives might be valid. In these cases, the pre-Darwinians arrived at erroneous conclusions because they were convinced that they had to make a choice between two processes which, in reality, occur simultaneously. For example, neither Lamarck nor Lyell understood speciation (the multiplication of species), but this failure led them to opposite conclusions. When looking at fossil faunas, Lamarck, a great believer in the adaptability of natural species, concluded that all the contained species must have evolved into very different descendants. Lyell, as an essentialist, rejected the possibility of a change in species and therefore he believed, like Cuvier, that all of the species had become extinct, with replacements provided by special creation. Neither Lamarck nor Lyell imagined that both processes, speciation and extinction, could occur simultaneously. That the turnover of faunas could be a balance of both processes never entered their minds.

FAILURE TO SEPARATE DISTINCT PHENOMENA

A third type of violation of scientific logic was particularly harmful to the acceptance of evolutionary thinking. This was the erroneous assumption that certain characteristics are inseparably combined. For instance, both Linnaeus and Darwin assumed, as I pointed out at an earlier occasion (*34*) that if one admitted the *reality* of species in nature, one would also have to postulate their immutable *fixity*. Lyell, as a good essentialist, unhesitatingly endorsed the same thesis: "From the above considerations, it appears that species have a real existence in nature; and that each was endowed, at the time of its creation, with the attributes and organization by which it is now distinguished" (*9,* p. 21). He is even more specific about this in his notebooks (*18,* p. 92). That species could have full "reality" in the nondimensional situation (*34*) and yet evolve continuously was unthinkable to him. Reality and constancy of species were to him inseparable attributes.

IMPACT OF THE ORIGIN OF SPECIES

The situation changed drastically and permanently with the publication of the *Origin of Species* in 1859. Darwin marshaled the evidence in favor of a transmutation of species so skillfully that from that point on the eventual acceptance of evolutionism was no longer in question. But he did more than that. In natural selection he proposed a mechanism that was far less vulnerable than any other previously proposed. The result was an entirely different concept of evolution. Instead of endorsing the 18th-century concept of a drive toward perfection, Darwin merely postulated change. He saw quite clearly that each species is forever being buffeted around by the capriciousness of the constantly changing environment. "Never use the word(s) higher and lower" *(35)* Darwin reminded himself. By chance this process of adaptation sometimes results in changes that can be interpreted as progress, but there is no intrinsic mechanism generating inevitable advance.

Virtually all the arguments of Cuvier, Lyell, and the progressionists became irrelevant overnight. Essentialism had been the major stumbling block, and the development of a new concept of species was the way to overcome this obstacle. Lyell himself eventually (after 1856) understood that the species problem was the crux of the whole problem of evolution, and that its solution had potentially the most far-reaching consequences: "The ordinary naturalist is not sufficiently aware that, when dogmatizing on what species are, he is grappling with the whole question of the organic world and its connection with a time past and with man" *(18,* p. 1). And, since he came to this conclusion after studying speciation in the Canary Islands, he added: "A group of islands, therefore, is the fittest place for Nature's trial of such permanent variety-making and where the problem of species-making may best be solved" *(18,* p. 93). This is what Darwin had discovered 20 years earlier.

SPECIAL ASPECTS OF THE DARWINIAN REVOLUTION

No matter how one defines a scientific revolution, the Darwinian revolution of 1859 will have to be included. Who would want to question that, by destroying the anthropocentric concept of the universe, it caused a greater upheaval in man's thinking than any other scientific advance since the rebirth of science in the Renaissance? And yet, in other ways, it does not fit at all the picture of a revolution. Or else, how could H. J. Muller have exclaimed as late as 1959: "One hundred years without Darwinism are enough!" *(36)*? And how could books such as Barzun's *Darwin, Marx, Wagner* (1941) and Himmelfarb's *Darwin and the Darwinian Revolution*

(1959), both displaying an abyss of ignorance and misunderstanding, have been published relatively recently? Why has this revolution in some ways made such extraordinarily slow headway?

A scientific revolution is supposedly characterized by the replacement of an old explanatory model by an incompatible new one *(1)*. In the case of the theory of evolution, the concept of an instantaneously created world was replaced by that of a slowly evolving world, with man being part of the evolutionary stream. Why did the full acceptance of the new explanation take so long? The reason is that this short description is incomplete, and therefore misleading, as far as the Darwinian revolution is concerned.

Before analyzing this more fully, the question of the date of the Darwinian revolution must be raised. That the year 1859 was a crucial one in its history is not questioned. Yet, this still leaves a great deal of leeway to interpretation. On one hand, one might assert that the age of evolutionism started even before Buffon, and that the publication of the *Origin* in 1859 was merely the last straw that broke the camel's back. On the other hand, one might go to the opposite extreme, and claim that not much had changed in the thinking of naturalists between the time of Ray and Tournefort and the year 1858, and that the publication of the *Origin* signified a drastic, almost violent revolution. The truth is somewhere near the middle; although there was a steady, and ever-increasing, groundswell of evolutionary ideas since the beginning of the 18th century, Darwin added so many new ideas (particularly an acceptable mechanism) that the year 1859 surely deserves the special attention it has received. Two components of the Darwinian revolution must thus be distinguished: the slow accumulation of evolutionary facts and theories since early in the 18th century, and the decisive contribution which Darwin made in 1859. Together these two components constitute the Darwinian revolution.

The long time span is due to the fact that not simply the acceptance of one new theory was involved, as in some other scientific revolutions, but of an entirely new conceptual world, consisting of numerous separate concepts and beliefs *(37)*. And not only were scientific theories involved, but also a whole set of metascientific credos. Let me prove my point by specifying the complex nature of the revolution: I distinguish six major elements in this revolution, but it is probable that additional ones should be recognized *(32)*.

The first three elements concern scientific replacements:

1. *Age of the earth*. The revolution began when it became obvious that the earth was very ancient rather than having been created only 6000 years ago *(17)*. This finding was the snowball that started the whole avalanche.

2. *Refutation of both catastrophism (progressionism) and of a steady-state world*. The evolutionists, from Lamarck on, had claimed that the concept of a more or less steadily evolving world, was in better agreement with the facts than either the

catastrophism of the progressionists or Lyell's particular version of a steady-state world. Darwin helped this contention of the evolutionists to its final victory.

3. *Refutation of the concept of an automatic upward evolution.* Every evolutionist before Darwin had taken it for granted that there was a steady progress of perfection in the living world. This belief was a straight-line continuation of the (static) concept of a scale of perfection, which was maintained even by the progressionists for whom each new creation represented a further advance in the plan of the Creator.

Darwin's conclusion, to some extent anticipated by Lamarck, was that evolutionary change through adaptation and specialization by no means necessitated continuous betterment. This view proved very unpopular, and is even today largely ignored by nonbiologists. This neglect is well illustrated by the teachings of the school of evolutionary anthropology, or those of Bergson and Teilhard de Chardin.

The last three elements concern metascientific consequences. The main reason why evolutionism, particularly in its Darwinian form, made such slow progress is that it was the replacement of one entire *weltanschauung* by a different one. This involved religion, philosophy, and humanism.

4. *The rejection of creationism.* Every antievolutionist prior to 1859 allowed for the intermittent, if not constant, interference by the Creator. The natural causes postulated by the evolutionists completely separated God from his creation, for all practical purposes. The new explanatory model replaced planned teleology by the haphazard process of natural selection. This required a new concept of God and a new basis for religion.

5. *The replacement of essentialism and nominalism by population thinking.* None of Darwin's new ideas was quite so revolutionary as the replacement of essentialism by population thinking (38, 19–23). It was this concept that made the introduction of natural selection possible. Because it is such a novel concept, its acceptance has been slow, particularly on the European continent and outside biology. Indeed, even today it has by no means universally replaced essentialism.

6. *The abolition of anthropocentrism.* Making man part of the evolutionary stream was particularly distasteful to the Victorians, and is still distasteful to many people.

NATURE OF THE DARWINIAN REVOLUTION

It is now clear why the Darwinian revolution is so different from all other scientific revolutions. It required not merely the replacement of one scientific theory by a new one, but, in fact, the rejection of at least six widely held basic beliefs [together with some methodological innovations (32)].

Furthermore, it had a far greater relevance outside of science than any of the revolutions in the physical sciences. Einstein's theory of relativity, or Heisenberg's of statistical prediction, could hardly have had any effect on anybody's personal beliefs. The Copernican revolution and Newton's world view required some revision of traditional beliefs. None of these physical theories, however, raised as many new questions concerning religion and ethics as did Darwin's theory of evolution through natural selection.

In a way, the publication of the *Origin* in 1859 was the midpoint of the so-called Darwinian revolution rather than its beginning. Stirrings of evolutionary thinking preceded the *Origin* by more than 100 years, reaching an earlier peak in Lamarck's *Philosophie Zoologique* in 1809. The final breakthrough in 1859 was the climax in a long process of erosion, which was not fully completed until 1883 when Weismann rejected the possibility of an inheritance of acquired characters.

As in any scientific revolution, some of the older opponents, such as Agassiz, never became converted. But the Darwinian revolution differed by the large number of workers who accepted only part of the package. Many zoologists, botanists, and paleontologists eventually accepted gradual evolution through natural causes, but not through natural selection. Indeed, on a worldwide basis, those who continued to reject natural selection as the prime cause of evolutionary change were probably well in the majority until the 1930s.

Two conclusions emerge from this analysis. First, the Darwinian and quite likely other scientific revolutions consist of the replacement of a considerable number of concepts. This requires a lengthy period of time, since the new concepts will not all be proposed simultaneously. Second, the mere summation of new concepts is not enough; it is their constellation that counts. Uniformitarianism, when combined with the belief in a static essentialistic world, leads to the steady-state concept of Lyell, while when combined with a concept of change, it leads to the evolutionism of Lamarck. The observation of evolutionary changes, combined with essentialist thinking, leads to various saltationist or progressionist theories, but, combined with population thinking, it leads to Darwin's theory of evolution by natural selection.

It is now evident that the Darwinian revolution does not conform to the simple model of a scientific revolution, as described, for instance, by T. S. Kuhn *(1)*. It is actually a complex movement that started nearly 250 years ago; its many major components were proposed at different times, and became victorious independently of each other. Even though a revolutionary climax occurred unquestionably in 1859, the gradual acceptance of evolutionism, with all of its ramifications, covered a period of nearly 250 years *(37)*.

References and Notes

1. T. S. Kuhn, *The Structure of Scientific Revolutions* (Univ. of Chicago Press, Chicago, 1962).

2. S. Toulmin, *Boston Stud. Phil. Sci.* **3**, 333 (1966); I. Lakatos and A. Musgrave, Eds., *Criticism and the Growth of Knowledge* (Cambridge Univ. Press, Cambridge, England, 1970), reviewed by D. Shapere, *Science* **172**, 706 (1971).

3. To cite only a few: L. Eiseley, *Darwin's Century* (Doubleday, New York, 1958); B. Glass, O. Temkin, W. L. Straus, Jr., Eds., *Forerunners of Darwin, 1745–1859* (Johns Hopkins Press, Baltimore, 1959); J. C. Greene, *The Death of Adam* (Iowa State Univ. Press, Ames, 1959); W. Zimmermann, *Evolution, Geschichte ihrer Probleme und Erkenntnisse* (Alber, Freiburg, West Germany, 1953); J. C. Greene, "The Kuhnian paradigm and the Darwinian revolution in natural history," in *Perspectives in the History of Science and Technology,* D. H. D. Roller, Ed. (Univ. of Oklahoma Press, Norman, 1971); M. T. Ghiselin, *New Lit. Hist.* **3**, 113 (1971).

4. G. de Beer, *Charles Darwin* (Doubleday, Garden City, N.Y., 1964).

5. C. C. Gillispie, *Genesis and Geology* (Harvard Univ. Press, Cambridge, Mass., 1951; rev. ed., Harper & Row, New York, 1959).

6. C. Darwin, *On the Origin of Species by Means of Natural Selection* (1859).

7. G. Cuvier, *Essay on the Theory of the Earth* (Edinburgh, ed. 3, 1817); much of it is an implicit refutation of Lamarck's ideas.

8. W. Coleman, *Georges Cuvier, Zoologist* (Harvard Univ. Press, Cambridge, Mass., 1964).

9. C. Lyell, *Principles of Geology* (John Murray, London, 1835), vols. 2 and 3 (I have used ed. 4). Important for British biology because it was the first presentation of Lamarck's theories to the English-speaking world (Book III, chap. I–XI). Darwin had previously heard about Lamarck from R. E. Grant in Edinburgh in 1827 [see *(4,* p. 28)].

10. L. Agassiz, *Essay on Classification* (Little, Brown, Boston, 1857; reprint, Belknap, Cambridge, Mass., 1962).

11. A. O. Lovejoy, "The argument for organic evolution before the *Origin of Species, 1830–1858,*" in *Forerunners of Darwin, 1745–1859,* B. Glass, O. Temkin, W. L. Straus, Jr., Eds. (Johns Hopkins Press, Baltimore, 1959), pp. 356–414.

12. A. O. Lovejoy, *The Great Chain of Being* (Harvard Univ. Press, Cambridge, Mass., 1963), Lecture 9.

13. The authorship of the anonymously published *Vestiges of the Natural History of Creation* (1844) did not become known until after the death of Robert Chambers [M. Millhauser, *Just Before Darwin* (Wesleyan Univ. Press, Middletown, Conn., 1959); see also *(5,* chap. 6)]. A sympathetic analysis of the *Vestiges* that does not concentrate on Chambers' errors and his gullibility, is still wanting [F. N. Egerton, *Stud. Hist. Phil. Sci.* **1**, 176 (1971)].

14. F. Darwin, Ed., *Life and Letters of Charles Darwin* (Sources of Science Ser. No. 102; reprint of 1888 ed., Johnson Reprints, New York, 1969), vol. 2.

15. H. Fruchtbaum, "Natural theology and the rise of science," thesis, Harvard University (1964).

16. A. Sedgwick, *Edinburgh Rev.* **82**, 3 (1845).

17. F. C. Haber, in *Forerunners of Darwin, 1745–1859,* B. Glass, O. Temkin, W. L. Straus, Jr., Eds. (Johns Hopkins Press, Baltimore, 1959), pp. 222–261.

18. L. G. Wilson, Ed., *Sir Charles Lyell's Scientific Journals on the Species Question* (Yale Univ. Press, New Haven, Conn., 1970).

19. K. R. Popper, *The Open Society and its Enemies* (Routledge & Kegan Paul, London, 1945).

20. D. Hull, *Brit. J. Phil. Sci.* **15**, 314 (1964); *ibid.* **16**, 1 (1965).

21. Aristotle is traditionally included among the essentialists, but newer researches cast considerable doubt on this. There is a growing suspicion that much of the late medieval thought labeled as Aristotelianism, had little to do with Aristotle's actual thinking. See, for instance, M. Delbrück, in *Of Microbes and Life,* J. Monod and E. Borek, Eds. (Columbia Univ. Press, New York, 1971), pp. 50–55.

22. E. Mayr, in *Evolution and Anthropology: A Centennial Appraisal* (Anthropological Society of Washington, Washington, D.C., 1959).

23. ———, *Harvard Libr. Bull.* **13**, 165 (1959).

24. O. L. Reiser, in *A Book that Shook the World,* R. Buchsbaum, Ed. (Univ. of Pittsburgh Press, Pittsburgh, Pa., 1958).

25. It is impossible in the limited space available to give a full documentation for the refutation of this thesis [see W. Coleman, *Biology in the Nineteenth Century* (Wiley, New York, 1971), p. 63].

26. G. Cuvier, *Essay on the Theory of the Earth,* R. Jameson, Transl. (Edinburgh, ed. 3, 1817). It is frequently stated that Cuvier believed in large-scale creations, necessary to repopulate the globe after major catastrophes, and this may well be true. However, I have been unable to find an unequivocal statement to this effect in Cuvier's writings [see also (*8,* p. 136)].

27. Progressionism was the curious theory according to which evolution did not take place in the organisms but rather in the mind of the Creator, who—after each catastrophic extinction—created a new fauna in the more advanced state to which His plan of creation had progressed in the meantime. This thought was promoted in Britain particularly by Hugh Miller (*Footprints,* 1847), Sedgwick (*Discourse,* 1850), and Murchison (*Siluria,* 1854), and in America by L. Agassiz (*Essay,* 1857); see (*3*).

28. The difference between catastrophism and uniformitarianism became smaller, as it was realized that many of the "catastrophes" had been rather minor events, and that contemporary geological phenomena (earthquakes, volcanic eruptions, tidal waves, glaciation) could have rather catastrophic effects [S. Toulmin, in *Criticism and the Growth of Knowledge,* I. Lakatos and A. Musgrave, Eds. (Cambridge Univ. Press, Cambridge, England, 1970), p. 42].

29. [W. Whewell] *Brit. Critic* **9**, 180 (1831); *Quart. Rev.* **47**, 103 (1832). For a full discussion of catastrophism see (5). A new interpretation of the traditional geological theories which Lyell opposed is given by M. J. S. Rudwick, "Uniformity and progression," in *Perspectives in the History of Science and Technology,* D. H. D. Roller, Ed. (Univ. of Oklahoma Press, Norman, 1971), pp. 209–237.

30. The term uniformitarianism was applied to at least four different concepts, and this caused considerable confusion, to put it mildly. For recent reviews see R. Hooykaas, *Natural Law and Divine Miracle* (E. J. Brill, Leiden, the Netherlands, 1959); S. J.

Gould, *Amer. J. Sci.* **263**, 223 (1965); C. C. Albritton, Jr., Ed., "Uniformity and simplicity," *Geol. Soc. Amer. Spec. Pap. No. 89* (1967); M. S. J. Rudwick, *Proc. Amer. Phil. Soc.* **111**, 272 (1967); G. G. Simpson, "Uniformitarianism," in *Essays in Evolution and Genetics* (Appleton-Century-Crofts, New York, 1970), pp. 43–96. The most important interpretations of uniformitarianism are: the same processes act now as in the past, the magnitude of geological events is as great now as in the past, and the earth a steady-state system.

31. Mrs. Lyell, Ed., *Life, Letters and Journals of Sir Charles Lyell* (John Murray, London, 1881), vol. 1, pp. 467–469 (letter of 1 June 1836 to J. W. Herschel).

32. Darwin's *Origin* was one of the first scientific treatises in which the hypothetico-deductive method was rather consistently employed [M. Ghiselin, *The Triumph of the Darwinian Method* (Univ. of California Press, Berkeley, 1969)]. Equally important, and even more novel, was Darwin's demonstration that deterministic prediction is not a necessary component of causality [M. Scriven, *Science* **130**, 477 (1959)]. Perhaps this can be considered a corollary of population thinking, but it is further evidence for the extraordinary complexity of the Darwinian revolution.

33. K. R. Popper, *The Logic of Scientific Discovery* (Hutchison, London, 1959).

34. E. Mayr, Ed., *The Species Problem* (AAAS, Washington, D.C., 1957), p. 2.

35. F. Darwin and A. C. Seward, Eds., *More Letters of Charles Darwin* (reprint of 1903 ed., Johnson Reprints, New York, 1971), vol. 1, p. 114.

36. H. J. Muller, "One hundred years without Darwinism are enough," *School Sci. Math.* 1959, 304 (1959).

37. It remains to be determined to what extent a similar claim can also be made for some of the physical sciences, for instance, the Copernican revolution.

38. In all recent discussions of natural selection, the assumption is made that the concept traces back to the tradition of Adam Smith, Malthus, and Ricardo, with the emphasis on competition and progress. This interpretation overlooks the point that the elimination of "degradations of the type" as the essentialists would call it, does not lead to progress. For the typologist, natural selection is merely the elimination of inferior types, an interpretation again revived by the mutationists (after 1900). Darwin was the first to see clearly that a second factor was necessary, the production of new variation. (This leads to population thinking.) Selection can be creative only when such new individual variation is abundantly available.

39. I greatly benefited from stimulating discussions with S. J. Gould and F. Sulloway, who read a draft of this essay, and from a series of most valuable critical comments, received from Prof. L. G. Wilson, which helped me to correct several errors. My own interpretation, however, still differs in some crucial points from that of Prof. Wilson. Some of the analysis was prepared while I served as Visiting Fellow at the Institute for Advanced Study, Princeton, N.J., in 1970.

The triune brain.

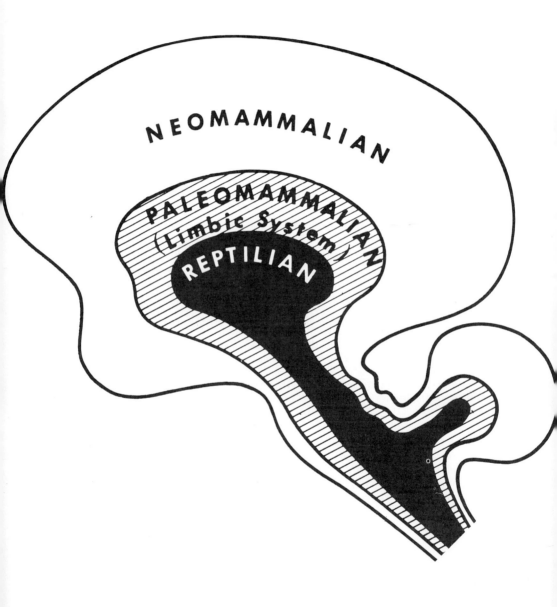

Paul D. MacLean

The Evolution of Three Mentalities

The irony of "objective attitudes" is that the cold hard facts of science are all derivatives of a "soft" brain. MacLean considers how the subjective brain affects our view of the world. His analysis of the brain and its functions depends on recognition of three levels of organization—the reptilian, paleomammalian, and neomammalian. These levels evolved millions of years apart and differ in structure and chemistry, but they are interconnected, each having its own special functions.

INTRODUCTION

Herein too may be felt the powerlessness of mere Logic, the insufficiency of the profoundest knowledge of the laws of the understanding, to resolve these problems which lie nearer to our hearts, as progressive years strip away from our life the illusions of its golden dawn.

George Boole, *An Investigation of the Laws of Thought*

Many people point out the apparent irony that the great strides in the natural sciences seem to be speeding us toward the Hill of Megiddo and the long-advertised final conflict between the forces of good and evil. Others, still blinded by the searing light of Hiroshima, are more introspective in expressing their concern. How, they ask, can we contain and harness the devastating powers of the atom before we have learned to understand and control the potentially catastrophic forces within ourselves?

In recent years anxiety about thermonuclear war has diminished somewhat in the light of warnings that the human race and many other forms of life may be on the way to extinction because of scientific developments that have made possible overpopulation, pollution of the environment, and exhaustion of critical resources.

A curve showing the growth of the world's population (cf. von Foerster et al. 1960) indicates that each successive doubling of people has taken place in half the time of the previous doubling (Calhoun 1971). At this rate the present population would be expected to double in 30 to 40 years. In 1969, U Thant, speaking as Secretary of the United Nations, made his famous pronouncement that there remained only 10 years to find solutions for the exploding population and related problems.

Warnings of this kind focus attention almost exclusively on the external environment. It is so easy to see the problems of meeting future demands for food, water, energy, and other basic requisites that planning experts seem to have overlooked the lessons of animal experimentation, which indicate that psychological stresses of crowding may bring about a collapse of social structure despite an ample provision of the necessities of life (e.g., Calhoun 1962; Myers et al. 1971). Systems analysts who have attempted to predict the limits of growth with the aid of computer technology (cf. Meadows et al. 1972) either admit to an inability to deal with psychological factors or neglect them altogether.

Michael Chance (1969) has remarked that the parts of the universe that man first chose to study were those farthest removed from the self—the heavens and the science of astronomy. Later I shall mention a possible neurological explanation of why our sciences from the very beginning have focused on the external world. Perhaps for similar reasons there has been a retarded interest in turning the dissecting lamp of the scientific method onto the inner self and the psychological instrument by which we derive all scientific knowledge. It would seem that there has always been a supernatural injunction against doing so: "Of every tree of the garden thou mayest freely eat: but of the tree of the knowledge of good and evil, thou

---

This article is from the introduction of a book (in preparation) on the triune brain. Not subject to copyright, having been prepared by an officer of the U.S. government.

shalt not eat of it: for in the day that thou eatest thereof thou shalt surely die."

Until recent times religion and philosophy have provided the principal interpreters of psychological matters. Although their modern origins were in the 18th century, psychology and psychiatry could hardly be regarded as sciences until the latter half of the 19th century. The same is true for neurophysiology and experimental psychology, which encompass investigations on the psychological functions of the brain. According to Kathleen Grange (1961), the term *psychology* was used in titles as early as 1703, while *psychiatry* first appeared on a title page in 1813. Psychiatry began to receive recognition as one of the medical sciences in 1854, when Griesinger at the University of Munich united for the first time the teaching of neurology and psychiatry. Meynert, Gudden, Forel, and others followed this practice and established it as a tradition in Europe. Since the middle of the present century neurology has followed an independent course, delving into psychological functions only insofar as particular disturbances in cerebration make it possible to diagnose the nature and location of brain disease. Psychoanalysis, which has given new conceptual and methodological dimensions to psychiatry, began to arouse public interest in 1900 with the publication of Freud's *The Interpretation of Dreams* (1900).

The late development of the psychological sciences is of itself of epistemological interest. This leads to the consideration that none of the psychological sciences devotes itself specifically to questions concerning the origin, nature, limits, and validity of knowledge. Except for sensation and perception, it is curious how little attention has been given by philosophers and others to the role of the brain in matters of epistemology.

Epistemology exists because of human societies, and human societies depend upon the existence of individuals. These truisms emphasize the incontrovertible centricity of the individual person with respect to public knowledge. In constitutional language, public knowledge—just as society itself—derives authority from individuals. In this sense an individual is both supreme and indispensable.

Central to every individual is a subjective self—a self that Descartes (1641) once referred to as "this me." A conceptual dissection of the subjective self requires that it be laid open not only in terms of its inner workings, but also in relationship to the societal and nonsocietal elements of the external environment. There are two sides to each of these relationships: the side that is intuitively and unsystematically experienced and the side that becomes known through the analytic and synthetic approaches of the various sciences. The animate relationships become systematically known through the social and life sciences, while formal knowledge of the inanimate derives from the natural sciences.

Epistemics

There is, however, no branch of science that deals specifically with an explanation of the subjective self and its relation to the internal and external environment. While such a study would draw upon every field of knowledge reflecting upon the human condition, it would build fundamentally on the psychological and brain-related sciences. In order to have a matching expression for epistemology, as well as an equivalent term for science, one might borrow a word directly from the Greek, and instead of speaking of a "science of the self," refer to an "episteme ($\epsilon\pi\iota\sigma\tau\eta\mu\eta$) of the self." Then the body of knowledge or the collective disciplines dealing with this subject could be known as *epistemics*.

Let it be emphasized that the domains of epistemics and epistemology are the same. The difference is in the point of view. Epistemics represents the subjective view and an epistemic approach from the inside out, whereas epistemology represents the public view and a scientific approach from the outside in. The two are inseparable insofar as epistemics is nuclear to epistemology and epistemology embraces epistemics. What is entailed is an obligatory relationship between a private, personal brain and a public, collective, societal brain.

Developments in the knowledge of the brain promise to have a profound influence on epistemology. In scientific and philosophic writings it has been customary to regard the human brain as a global organ dominated by the cerebral cortex, which serves as a *tabula rasa* for an ever-changing translation of sensory and perceptive experience into symbolic language and which has special capacities for learning, memory, problem solving, and the transmission of culture from one generation to another. Such a view is blind to the consideration that in its evolution the human brain has expanded according to three basic patterns, which may be characterized as reptilian, paleomammalian, and neomammalian (Figure 1) (MacLean 1962). Radically different in structure and chemistry, and in an evolutionary sense countless generations apart, the three formations constitute three brains in one, a *triune* brain (MacLean 1970, 1973c). What this situation immediately implies is that we are obliged to look at ourselves and the world through the eyes of three quite different mentalities. To complicate things further, two of the mentalities appear to lack the power of speech.

Objectivity*

Achievements of the "hard" or exact sciences have helped to promote the attitude that solutions to most problems can be found by learning to

---

*The wording of this passage follows closely that of MacLean 1970.

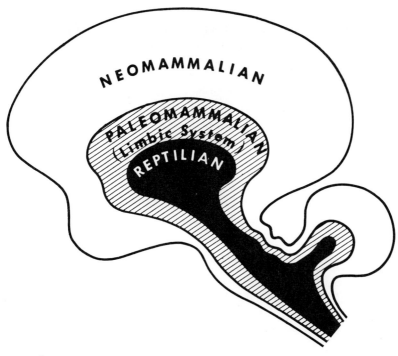

FIGURE 1
In its evolution, the human forebrain expands in hierarchic fashion along the lines of three basic patterns that may be characterized as reptilian, paleomammalian, and neomammalian (MacLean 1967).

manipulate the external environment. It has been traditional to regard the exact sciences as completely objective. The self-conscious cultivation of the "objective" approach is illustrated in a statement by Einstein, quoted by C. P. Snow (1967: 90): "A perception of this world by thought, leaving out everything subjective became . . . my supreme aim." In a recent essay on the contributions of molecular biology, Monod (1971) is equally insistent on applying the "principle of objectivity" in the life sciences. "The cornerstone of the scientific method," he writes, "is the postulate that nature is objective" (p. 21). Even in the world of fiction one finds a book reviewer saying, "Humanity is likely to be saved, if it is at all, by a search for an objective reality we can all share—for truths like those of science" (Weisberger 1972).

37

Early in this century, John B. Watson and others of the behaviorist school sought to revive the spirit of the Helmholz tradition by establishing psychology as an exact science on an equal footing with physics and the other natural sciences (cf. Shakow and Rapaport 1964). In their study of animals and man these theorists advocated a completely objective approach that dispensed with the consideration of consciousness, subjectivity, and introspection (Watson 1924). The irony of all such objective attitudes is that every behavior selected for study, every observation and interpretation, requires subjective processing by an introspective observer. Logically, there is no way of circumventing this or the more disturbing conclusion that the cold hard facts of science, like the firm pavement underfoot, are all derivatives of a "soft" brain. No measurement or computation obtained by the hardware of the exact sciences enters our comprehension without undergoing subjective transformation by the software of the brain. The implication of Spencer's statement (1896) that objective psychology owes its origins to subjective psychology could apply to the whole realm of science.

For such reasons it is important to consider how a fifth dimension, the subjective brain, affects our relative view of the world. In considering this problem I do not intend to deal with the familiar Cartesian topic of perceptual illusions. Rather, I will focus on brain research concerned with the origins of other forms of experience and attitudes that may be of more basic significance for epistemics and epistemology, giving particular attention to forebrain mechanisms underlying *paleopsychic* processes and *prosematic* (non-verbal) behavior.

Subjective Experience

For each of us as individuals there is nothing so vital as our subjective experience. Without the essence of subjectivity, there would be no means of realizing our existence. Subjectivity represents a form of information. As Wiener (1948) stated more succinctly than Berkeley or Hume, "Information is information, not matter or energy" (p. 155). At the same time, it is empirically evident that there can be no communication of information without the intermediary of what we recognize as physical behaving entities. This invariance might be considered a law of communication.

Facts

I should also mention at this point that facts apply only to those things that can be agreed upon publicly as entities behaving in a certain way. The term *validity* does not apply to the facts themselves, which are neither true nor false per se, but rather to what is agreed upon as true by subjective individuals after a public assessment of the facts. What is agreed upon as true or false by one group may be quite contrary to the conclusions of another group.

Communicative Behavior

Next in importance to our subjective experience is our ability to share what we feel and think with other beings. Such communication must be accomplished through some form of behavior. Human communicative behavior can be broadly categorized as verbal and nonverbal. Like P. W. Bridgman (1959), the physicist-philosopher, the great majority of people would probably conclude that most communication is verbal. Since we are accustomed to thinking of ourselves as verbal beings, we have given less attention to the analysis of nonverbal communication. This neglect is evidenced by our lack of a specific word for such nonverbal behavior; we refer to it negatively by stating what it is not.

It is an everyday experience that in spite of all kinds of talk—no matter how well documented—we are never quite sure how we develop attitudes or reach decisions regarding all manner of human relationships. Who would feel confident in trying to identify the nonverbal factors affecting one's choice of spouse, friends, associates; a vote for a particular candidate; one's judgments as a member of a committee or jury? In an article on nonverbal communication in Japan, Morsbach (1973) illustrates the bewilderment commonly felt in trying to reconstruct human decisions. In an anecdote conjuring a feeling of *déjà vu* he describes two professors who after a faculty meeting found themselves in agreement that everyone had spoken positively about a particular proposal that was subsequently voted down. "Don't you agree," one asked the other, "that everyone was in favor?" "Yes," was the reply, "but you did not hear the silences."

Contrary to the popular view, many behavioral scientists would be inclined to give greater importance to nonverbal than verbal behavior in day-to-day human activities. For example, when a psychologist, a behavioral ecologist, a specialist in environmental design, and an ethologist were asked to draw two squares representing the weight that they would give verbal and nonverbal communication in everyday human activity, there was a striking similarity in their responses. In each case, the square for nonverbal behavior was about three times the size of the one for verbal behavior. It must be admitted, however, that we are so ignorant of the hidden aspects of nonverbal behavior that it would be impossible to make quantitative assessments of their influence.

Nonverbal (Prosematic) Behavior

Nonverbal behavior mirrors in part what Freud (1900) called primary processes. In drawing a distinction between verbal and nonverbal behavior it is easier to see differences than similarities. But in a very real sense, nonverbal behavior, like verbal behavior, has its semantics and syntax—in other words, *meaning* and *orderly arrangement* of specific acts.

It is nonverbal behavior that we possess in common with animals. Since it is hardly appropriate to refer to *nonverbal* behavior of animals (cf. Hinde 1972), it is desirable to use some other term for this kind of behavior. The Greek word σημα pertains to a sign, mark, or token. By adding the prefix προ in the particular sense of "rudimentary" one obtains the word *prosematic,* which would be appropriate for referring to any kind of nonverbal signal communication—vocal, bodily, chemical (MacLean 1974, 1976).

It has been the special contribution of ethology to provide the first systematic insights into the semantics and syntax of animal behavior (Lorenz 1937; Tinbergen 1951). An analysis of prosematic behavior of animals reveals that somewhat analogous to words, sentences, and paragraphs, it becomes meaningful in terms of its components, constructs, and sequences of constructs. Since the patterns of behavior involved in self-preservation and survival of the species are generally similar in most terrestrial vertebrates, it is inappropriate to speak, as in the past, of species-specific behavior. But since various species perform these behaviors in their own typical ways, it is both correct and useful to refer to species-typical behavior.

Introspectively, we recognize that prosematic communication may be either active or passive. When two or more individuals are within communicative distance there is the possibility for either active (intentional) or passive (unintentional) communication to occur with respect to the sender or receiver. Even when an individual is alone, a sound, utterance, movement, or odor emanating from the self may have self-communicative value as it originates either actively or passively.

SYNOPSIS OF EXPERIMENTAL WORK

For the past twenty-five years my research has been primarily concerned with identifying and analyzing forebrain mechanisms underlying prosematic forms of behavior, which on phylogenetic and clinical grounds might be inferred to represent expressions of *paleopsychic* processes. In this work, I have taken a comparative evolutionary approach that has the advantage of allowing one to telescope millions of years into a comprehensive span making it possible to see trends that would not otherwise be apparent. It also shows the usefulness of research on animals for obtaining insights into brain mechanisms underlying human prosematic behavior.

Since animal experimentation provides our only systematic knowledge of brain functions, I should comment briefly upon the justification of using findings on animals for drawing inferences about the workings of the human brain. There is general enthusiasm for applying findings on animals to human biology at the molecular and cellular levels. In the field of psychiatry, neurochemical and neuropharmacological discoveries in animals have radically changed the treatment of certain neuropsychiatric disorders. But many people believe that behavioral and neurological observations on animals have little or no human relevance.

Perhaps such a bias stems from a failure to realize that in its evolution, the human brain expands in hierarchical fashion along the lines of three basic patterns (reptilian, paleomammalian, and neomammalian (Figure 1). These three formations are distinctly different in chemistry and structure and in an evolutionary sense eons apart. Extensively interconnected, the three basic formations represent an amalgamation of three brains in one, or what may be appropriately called a triune brain (MacLean 1970, 1973b, 1973c). The word *triune* also implies that the whole is greater than the sum of its parts, because with the exchange of information among the three formations, each derives a greater amount of information than if it were operating alone. Stated in popular terms, the amalgamation amounts to three interconnected biological computers, each having its own special intelligence, its own subjectivity, its own sense of time and space, its own memory, and its own motor and other functions.

This scheme for subdividing the brain may seem simplistic, but thanks to improved anatomical, physiological, and chemical techniques, the three basic formations stand out in clearer detail than ever before. Moreover, it should be emphasized that despite their extensive interconnections, these brain types are capable of operating somewhat independently. Most important in regard to the verbal-nonverbal question, there are clinical indications that the reptilian and paleomammalian formations lack the neural machinery for verbal communication. To say that they lack the power of speech, however, does not belittle their intelligence, nor does it relegate them subjectively to the realm of the unconscious.

The basic neural machinery required for self-preservation and the preservation of the species is built into the neural chassis contained in the midbrain, pons, medulla, and spinal cord. As shown by the early experiments of Ferrier (1876) and others, an animal with only its neural chassis is as motionless and aimless as an idling vehicle without a driver. But with the evolution of the forebrain, the neural chassis acquires three drivers, all of different minds and all vying for control.

41

The Reptilian-type Brain

Let us look first at the reptilian "driver." In mammals, the major counterpart of the reptilian forebrain is represented by a group of large ganglia including the olfacto-striatum, corpus striatum (caudate nucleus and putamen), globus pallidus, and satellite gray matter. Since there is no name that applies to all of these structures, I shall refer to them as the R-complex. As shown in Figure 2, the stain for cholinesterase reveals a remarkable chemical contrast between the R-complex and the two other cerebrotypes. The shaded areas in Figure 3 show how this stain sharply demarcates the R-complex in animals ranging from reptiles to man. In using the fluorescent technique of Falck and Hillarp (1959), it is striking to see how the structures corresponding to those in the figure glow a bright green because of large amounts of dopamine (Juorio and Vogt 1967).

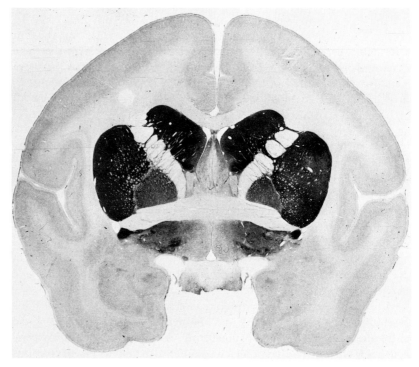

FIGURE 2
This section from the brain of a squirrel monkey shows how the greater part of the R-complex is selectively colored (black areas) by a stain for cholinesterase (MacLean 1972a).

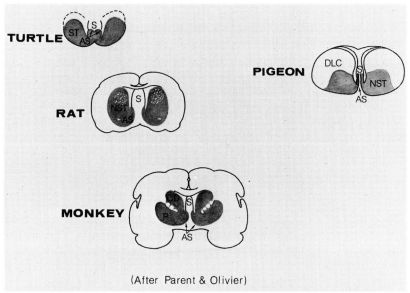

(After Parent & Olivier)

FIGURE 3

Shaded areas indicate how a stain for cholinesterase demarcates the greater part of the R-complex in animals ranging from reptiles to primates. With the fluorescent technique of Falck and Hillarp, the same areas shown above would glow a bright green because of the high content of dopamine. The pallidal part of the striatal complex does not fluoresce. No existing reptiles represent the forerunners of mammals. Birds are an offshoot from the *Archosauria* ("ruling reptiles") (after Parent and Olivier 1970).

From an evolutionary standpoint it is curious that ethologists have paid little attention to reptiles, focusing instead on fishes and birds. Some authorities believe that of the existing reptiles, lizards would bear the closest resemblance to the mammal-like reptiles believed to be the forerunners of mammals. At all events, lizards and other reptiles provide illustrations of complex prototypical *patterns of behavior* commonly seen in mammals, including man. One can quickly list more than 20 such behaviors that may primarily involve self-preservation or the survival of the species (MacLean 1976): (1) selection and preparation of homesite; (2) establishment of domain or territory; (3) trail making; (4) marking of domain or territory; (5) showing place preferences; (6) ritualistic display in defense of territory, commonly involving the use of coloration and adornments; (7) formalized intraspecific fighting in defense of territory; (8) triumphal display in successful defense; (9) assumption of distinctive postures and coloration in signaling surrender; (10) routinization of daily activities; (11) foraging; (12) hunting; (13) homing; (14) hoarding; (15) use of defecation posts; (16) formation of social groups; (17) establishment of social hierarchy by ritualistic display and other means; (18) greeting; (19) grooming; (20) courtship, with displays using coloration and adornments; (21) mating; (22) breeding and, in isolated instances, attending offspring; (23) flocking; and (24) migration.

Five Interoperative Behaviors

There is an important *pentad* of prototypical forms of behavior of a general nature that may be variously operative in the above activities. They may be denoted as (1) isopraxic, (2) perseverative, (3) reenactment, (4) tropistic, and (5) deceptive behavior. The word *isopraxic* will be used to refer to behaviors in which two or more individuals engage in the same kind of activity. Purely descriptive, it avoids preconceptions and prejudices commonly attached to such terms as *social facilitation* and *imitation.* Perseverative behavior applies to repetitious acts, such as occur in displays or in conflictive situations. Reenactment behavior refers to the repetition on different occasions of behaviors seeming to represent obeisance to precedent as, for example, following familiar trails or returning year after year to the same breeding grounds. Tropistic behavior is characterized by positive or negative responses to partial or complete representations of animate or inanimate objects and includes what ethologists refer to as imprinting and fixed-action patterns. Deceptive behavior involves the use of artifice and deceitful tactics such as are employed in stalking a prey or evading a predator. Except for altruistic behavior and most aspects of parental behavior, it is remarkable how many patterns of behavior seen in reptiles are also found in human beings.

As yet, hardly any investigations have been conducted on reptiles in an attempt to identify specific structures of the forebrain involved in the various behaviors listed above. All that is known thus far is that the neural guiding systems for species-typical, complex forms of behavior lie forward of the neural chassis.

In contrast to reptiles, the R-complex of mammals has been subjected to extensive investigation. Curiously enough, however, 150 years of experimentation have revealed remarkably little about its functions. The finding that large destructions of the mammalian R-complex may result in no obvious impairment of movement speaks against the traditional clinical view that it subserves purely motor functions. At the Laboratory of Brain

Evolution and Behavior of the National Institute of Mental Health, we are conducting comparative studies of reptiles, birds, and mammals in which we are testing the hypothesis that the R-complex plays a basic role in species-typical, prosematic behavior.

So far, crucial findings relevant to prosematic behavior have developed from experiments on more than 100 squirrel monkeys *(Saimiri sciureus)*. Animals of this species perform a characteristic display of the erect phallus in a show of aggression, in courtship, and as a form of greeting (MacLean 1962; Ploog and MacLean 1963). Members of one subspecies consistently display to their reflections in a mirror, providing observers a means of systematically testing the effects of brain ablations on the incidence and manifestations of the display (MacLean 1964). I have found that large bilateral lesions of the paleo- and neomammalian parts of the forebrain may have either no effect or only a transitory effect on the display. After bilateral lesions of the pallidal part of the R-complex (MacLean 1972, 1973a), however, or interruption of its main pathways (MacLean 1975), monkeys may no longer show an inclination to display. Without a test of the innate display behavior, one might conclude that they were unaffected by the loss of brain tissue.

These experiments provide the first evidence in mammals that the R-complex and its major pathways are basically involved in the performance of genetically constituted, species-typical, prosematic behavior. Such work represents a necessary first step for a more detailed analysis of forebrain mechanisms underlying territorial assertiveness, courtship, and social deportment. Since the mirror display also involves isopraxic factors, the experiments indicate that the R-complex is implicated in *natural* forms of imitation.

The Paleomammalian Brain

There are behavioral indications that the reptilian brain is poorly equipped for learning to cope with new situations. The reptilian brain has only a rudimentary cortex. In the lost transitional forms between reptiles and mammals—the mammal-like reptiles—it is presumed that the primitive cortex underwent further elaboration and differentiation. The primitive cortex is comparable to a crude radar screen, providing the animal with a better means of viewing the environment and learning to survive. In all existing mammals the phylogenetically old cortex is found in a large convolution which the 19th-century anatomist Broca (1878) called the great limbic lobe because it surrounds the brain stem. (Limbic means "forming a border around.") As illustrated in Figure 4, the limbic lobe forms a common denominator in the brains of all mammals. In 1952, I suggested the

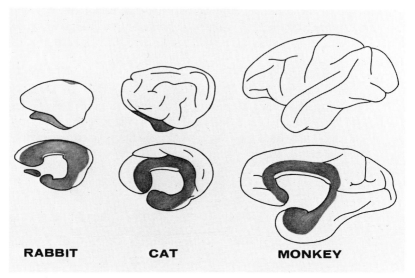

RABBIT   CAT   MONKEY

FIGURE 4

The limbic lobe of Broca (shaded) is found as a common denominator in the brains of all mammals. It contains the greater part of the cortex representative of the paleomammalian brain. The cortex of the neomammalian brain (shown in white) mushrooms late in evolution (MacLean 1954).

term *limbic system* as a designation for the limbic cortex and structures of the brain stem with which it has primary connections.

The limbic cortex is structurally less complicated than the new cortex. It was once believed to receive information mainly from the olfactory and visceral systems. However, we have shown, by recording from single nerve cells in awake, sitting monkeys, that signals also reach it from the visual, auditory, and somatic senses. There are clinical indications that the combined reception of information from the inside and outside worlds is essential for a feeling of individuality and personal identity (MacLean 1972b).

Also in contrast to the new cortex, the limbic cortex has large cablelike connections with the hypothalamus, which has long been recognized to play a central role in integrating the performance of mechanisms involved in self-preservation and the procreation of the species.

Although the limbic system undergoes considerable expansion in the brains of higher mammals, the basic pattern of organization remains the same as in lower mammals. Electrophysiological studies have shown that this basically paleomammalian brain is functionally, as well as anatomically, an integrated system. In the past 40 years clinical and experimental investigations have provided evidence that the limbic system derives information in terms of emotional feelings that guide behavior with respect to the two basic life principles of self-preservation and the preservation of the species.

Before further comment on limbic functions, it should be noted that many people maintain that it is inadmissible to make a sharp distinction between emotion and reason. Raphael Demos (1937), in an introduction to the dialogues of Plato, expresses a traditional philosophical view: ". . . we are apt to separate reason from emotion. Plato does not. Reason is not merely detached understanding; it is conviction, fired with enthusiasm" (p. xi). Piaget (1967), the founder of the Center for Genetic Epistemology, is quite vehement, saying that "nothing could be more false or superficial" than to attempt "to dichotomize the life of the mind into emotion and thoughts. . . . Affectivity and intelligence," he insists, "are indissociable and constitute the two complementary aspects of all human behavior" (p. 15).

Given the complementary aspects of emotion and thought, evidence from the study of psychomotor epilepsy shows that the two may occur independently because they are products of different cerebral mechanisms. Clinical observations provide the best evidence of the role of the limbic system in emotional behavior. Epileptic discharges in or near the limbic cortex result in a broad spectrum of vivid emotional feelings. It is one of the wonders of the brain that limbic discharges tend to spread in and be confined to the limbic system, not directly involving the neocortex. I have referred to this condition as a *schizophysiology* (1954) and have suggested that the underlying factors may contribute to inexplicable conflicts between "what we feel" and "what we know."

In regard to structures possibly involved in mental illness, it is significant that limbic discharges may result in symptoms characteristic of the toxic and endogenous psychoses, such as feelings of depersonalization, distortions of perception, paranoid delusions, and hallucinations (MacLean 1973c). I referred earlier to the striking chemical differences of the three basic cerebrotypes. An accumulation of evidence indicates that many of the psychotherapeutic drugs owe their salutary effects to a selective action on the limbic system and the R-complex.

It is of special epistemological interest that at the beginning of a limbic discharge, a patient may have intense free-floating feelings about what is real, true, and important or experience eureka-type feelings like those associated with discovery. There may be oceanic feelings such as occur in a mystical revelation or under the influence of psychedelic drugs. Ironically, it seems that the ancient limbic system has the capacity to generate strong affective feelings of conviction that we attach to our beliefs, regardless of whether they are true or false!

Three Subdivisions of the Paleomammalian Brain

The limbic system comprises three subdivisions (MacLean 1958). The two older ones (Figure 5) are closely related to the olfactory apparatus. Our experimental work has provided evidence that these two divisions are involved respectively in oral and genital functions. The findings are relevant to orosexual manifestations in feeding situations, mating, and in aggressive behavior and violence. The close relationship between oral and genital functions seems to be due to the olfactory sense which, dating far back in evolution, is involved in both feeding and mating.

The main pathway to the third subdivision bypasses the olfactory apparatus. In evolution, this subdivision reaches its greatest development in the human brain. An assortment of evidence suggests that this remarkable expansion reflects a shift from olfactory to visual and other influences in sociosexual behavior. It is also possible that this subdivision, together with the prefrontal cortex of the neomammalian brain, has provided a neural substrate for the evolution of human empathy.

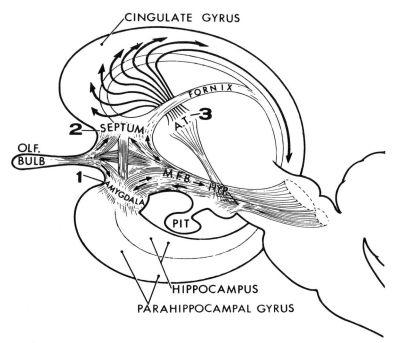

FIGURE 5

Diagram of three main subdivisions of the limbic system and their major pathways. See text for summary of their respective functions. Abbreviations: AT, anterior thalamic nuclei; HYP, hypothalamus; MFB, medial forebrain bundle; PIT, pituitary; OLF, olfactory (MacLean, 1958).

Avenues to the Basic Personality

The major pathways to and from the reptilian-type and paleomammalian-type brains pass through the hypothalamus and subthalamic region. If the majority of these pathways are destroyed in monkeys, the animals are greatly incapacitated, but with careful nursing they may recover the ability to feed themselves and move around. The most striking characteristic of these animals is that although they look like monkeys, they no longer behave like monkeys. Almost everything characteristic of species-typical, simian behavior has disappeared. If one were to interpret these experimental findings in the light of certain clinical case material, one might say that these large connecting pathways between the reptilian and paleomammalian formations provide the avenues to the basic personality. Here, certainly, would seem to be the pathways to the expression of prosematic behavior.

The Neomammalian Brain

Compared with the limbic cortex, the neocortex (shown in white in Figure 4) is like an expanding numerator. As C. Judson Herrick (1933) has commented, "Its explosive growth late in phylogeny is one of the most dramatic cases of evolutionary transformations known to comparative anatomy." The massive proportions achieved by the neocortex in higher mammals explain the designation of neomammalian brain applied to it and to the structures of the brain stem with which it is primarily connected. The neocortex culminates in the human brain, affording a vast neural screen for the portrayal of symbolic language and the associated functions of reading, writing, and arithmetic. Mother of invention and father of abstract thought, it promotes the preservation and procreation of ideas (MacLean 1973b). As opposed to the limbic cortex, the sensory systems projecting to the neocortex primarily give information about the external environment—namely, the visual, auditory, and somatic systems. It therefore seems that the neocortex is primarily oriented toward the outside world. Here, perhaps, is a clue to what was mentioned earlier regarding the traditional emphasis of the sciences on the external environment.

Three Forms of Mentation

A brief discussion of questions relevant to brain research will require me to use the expressions protomentation, emotional mentation, and rational mentation. Protomentation applies to rudimentary mental processes underlying complex prototypical forms of behavior, as well as the propensions mentioned above. Emotional mentation refers to cerebral processes underlying what are popularly regarded as emotions. Paleopsychic processes cover those aspects of protomentation and emotional mentation that are manifest by prosematic behavior. The meaning of rational mentation (ratiocination) will be assumed to be self-evident.

QUESTIONS RELEVANT TO BRAIN RESEARCH

In the past there has been the tendency to lump together many of the psychological processes that have just been alluded to in regard to protomentation and emotional mentation. Originally the German word *Trieb,* which Freud (1900) used to refer to drive, impulse, or urge, was inappropriately translated into English as "instinct." This twist led to the commonly used expression "instinctual drives." Instincts were regarded as the biological driving forces that, analogous to the pressure head in a hydraulic system, impelled an individual to action. The pressure of the instinctive forces was believed to result in emotional feelings of an unpleasant sort, while the reduction of tension gave rise to pleasurable feelings. Freud used the impersonal word *id* ("it") to apply to the instinctual forces of the so-called "unconscious" part of the "mental personality." In 1933, at the age of 77, he stated in *New Introductory Lectures,* "In popular language . . . we may say that the id stands for the untamed passions" (1949: 102).

Since the disposition to equate the instincts and emotions has continued to the present day, it will be timely to point out developments in the study of the brain and behavior that bring into better focus some distinctions between protomentation and emotional mentation. In addition to the role of protomentation in special prototypical behaviors such as are observed in the establishment of territory, we shall give consideration to the part it plays in the pentad of general prototypical behaviors as they become manifest in obsessive-compulsive behavior: day-to-day rituals of which we are hardly aware; the tendency to seek and give obeisance to precedent as in legal and other matters; superstitious actions; deceptive behavior; and imitation (isopraxis).

Because of their subjective obtrusiveness, perhaps a disproportionate emphasis has been given to the importance of emotions in influencing our day-to-day activities, but we should keep in mind the possibility that emotions may oftentimes be passive reflectors of psychic states rather than determinants of action and that under many circumstances propense forms of protomentation may play a more basic role.

The world's literature provides abundant evidence that because of moral customs and society's numerous ways of meting out punishment for wrongdoing, we tend to give greater weight to the role of emotions in unpleasant affairs than in situations of joy and gratification. This biased attitude carries over into scientific literature where one finds an authority on brain mechanisms of emotion referring to anger and fear as the "major emotions" (Bard 1928).

Some Medical and Legal Aspects

The negative role assigned to the emotions in medical and legal matters has a long history. Grange (1961) points out that in the 18th century "moral" insanity was the equivalent of emotional insanity. In an article on Pinel, the well-known 18th-century psychiatrist, she explains how he helped to popularize the use of the word *moral* to describe emotional factors in mental experience. He believed that the chief cause of insanity was moral, and his treatment of insanity was based on Aristotle's theory of "balancing the passions." Grange also describes how Pinel's writings inspired others to examine the emotions "in relation to health, to education, to politics, to crime, to urban and rural environments, to organic disease, and to the welfare of groups and individuals" (1961: 452). Today, just as two centuries ago, emotions are commonly believed to be the root of psychoneuroses, several forms of psychoses, psychosomatic disease, alcoholism, narcotic addiction, implacable domestic situations, juvenile delinquency, and crime. But developing insights require us to keep in mind that protomentation may be more fundamentally involved than emotional mentation, recalling, symbolically, that "the reptile does what it has to do."

There is another side of the emotions that has received less attention, possibly because it is considered an unalterable part of human nature. I refer now to the paradoxical capacity of emotional mentation to find support for opposite sides of any question. We take it for granted that the emotions will generate divisiveness in every field of discourse whether it is religious, ethical, artistic, sociological, legal, economic, political, educational, or scientific. However much this may be lamented, there is the potential benefit that argument and conflict may stir up the "gene pool" of ideas and lead to new and constructive concepts.

All such concerns about the emotions seem insignificant, however, compared with those deeply personalized feelings experienced when the utter isolation of death separates us from a loved one, or finally, after long prospect, forces itself upon each one of us.

Some Scientific Implications

After the presentation of behavioral and experimental data, it will be of interest to consider how in scientific affairs protomentation and protoreptilian propensities may be influential in regard to the establishment of intellectual domain (territory), the *idée fixe* of such a scientist as Kepler, obeisance to precedent, adherence to doctrine, and intolerance of new ideas.

As regards emotional mentation, it seems a particular irony that in science, as in politics, the "emotions" make it possible to stand on any platform. How does it happen that different groups of reputable scientists presented with the same data often find themselves at opposite poles—and sometimes in bitter, acrimonious debate—because of diametrically opposed views of what is true? It is equally puzzling—and intellectually incongruous—that for centuries the world order of science may emotionally cling to, and champion, "suspect" beliefs that are destined to crumble. What makes it psychologically possible for wise men to build higher and higher on these foundations without fear of their sudden collapse? It is also curious that the emotional investment of some scientists is so great that they remain convinced of the truth of a theory long after it has proven to be false. As the late E. G. Boring (1964) commented in paraphrasing a statement of Max Planck, "Important theories, marked for death by the discovery of contradictory evidence, seldom die before their authors" (p. 682). Two recent essays (Washburn 1973; Washburn and Ciochon 1974) have presented an instructive analysis of how emotional factors seem to have been instrumental in solidifying the thinking of proponents of divergent views of human evolution despite the admitted lack of sufficient data. They suggest some insightful correctives.

Why Brain Research?

Why is it so necessary to investigate brain mechanisms in order to understand the various forms of paleopsychic processes under consideration? After all, the laws of formal thought have been derived without an understanding of the underlying machinery of the brain. It is the peculiarity of rational mentation that it lends itself, as in the case of logic, to symbolic representation in the form of words or other signs which, when semantically specified and syntactically related according to certain rules, result in inevitable conclusions. A parallel situation applies to numerical procedures in which the steps of calculation are interlocked, assuring an outcome as predictable as the movements of a gear train. Within a generation, we have seen the evolution from simple calculating machines to giant computers which, when programmed according to the laws of logic, can reach the solution of a problem in a fraction of the time formerly required.

In formal rational mentation we have the advantage of being able to specify the inputs into our own brains or the prosthetic brains of computers. But the situation is quite different in the case of protomentation and emotional mentation. Here the known input is so obscured by an indefinable input from the person's ancestral past and personal life history that there is no means of determining the outcome. The successive mentational processes have neither been identified nor shown to obey laws that allow predictable conclusions.

Because of the inability to specify and control the internal input for paleopsychic processes, there is the hope that insights may be gained from an investigation of underlying mechanisms. Until the restrictions of the mechanisms are known, there can be as many explanations of paleopsychic processes as there are explicators.

Many learning theorists and behaviorists would take exception to this point of view, contending that the cranium and its contents may be regarded as a black box. Some adherents to operant conditioning claim that in utilizing the principle of reinforcement, any form of behavior can be shaped and predicted. Until it can be demonstrated, however, that neuroses in animals and human beings can be regularly induced and alleviated by operant techniques, one may reserve judgment about such claims.

Jeans (1943) has stated that "physics gives us exact knowledge because it is based on exact measurements." But if the ultimate scientific instrument, the human brain, is for one reason or another predisposed to artifactual interpretations, where does confidence lie in any field?

Time and Space

Although nuclear physicists are quick to point out the evaporation of the material world at the atomic level, many of them seem to retain an abiding faith in the existence of time and space. They would contend that if all particles were to disappear from the universe, space and time would still remain. There is an evident inconsistency in such an argument when we look at what Kant said about the "transcendental aesthetic." In view of the tripartite division of the brain under consideration, we want to keep in mind the question of whether there exist "reptilian time," "paleomammalian time," and "neomammalian time." A parallel question applies to space. Recently, students of environmental design have begun to consider the latter question in connection with urban planning and the desirable uses of space (Mallows 1970; Esser 1974; Greenbie 1975).

CONCLUDING COMMENTS

These introductory remarks on the evolution of three mentalities have implied that the reptilian, paleomammalian, and neomammalian formations provide three underlying neural mechanisms for what have been provisionally referred to as protomentation, emotional mentation, and rational mentation. At the same time, I have used the expression *triune brain* to indicate that no hard and fast boundaries exist between the three formations and their respective functions. With these provisions, I use a metaphor to summarize.

In the field of literature an irreducible number of basic plots and associated emotions are recognized. In describing the functions of the triune brain metaphorically, one might imagine that the reptilian brain provides the basic plots and actions; that the limbic brain influences emotionally the developments of the plots; while the neomammalian brain has the capacity to expound the plots and emotions in as many ways as there are authors. Such, in a capsule, is the theme underlying this chapter.

References Cited

Bard, P.
    1928   A diencephalic mechanism for the expression of rage with special reference to the sympathetic nervous system. Amer. J. Physiol. 84: 490–513.

Boring, E. G.
    1964   Cognitive dissonance: its use in science. Science 145: 680–685.

Bridgman, P. W.
    1959   The Way Things Are. Cambridge, Mass.: Harvard University Press.

Broca, P.
    1878   Anatomie comparée des circonvolutions cérébrales. Le grand lobe limbique et la scissure limbique dans la série des mammifères. Rev. Anthrop. 1: 385–498.

Calhoun, J. B.
    1962   Population density and social pathology. Sci. Amer. 206: 139–146.

    1971   Space and the strategy of life. *In* Behavior and Environment. A. H. Esser, ed. Pp. 329–387. New York: Plenum Press.

Chance, M.
    1969   Towards the biological definition of ethics. *In* Biology and Ethics. J. Ebling, ed. Pp. 3–13. New York and London: Academic Press.

Demos, R.
    1937   Introduction: The Dialogues of Plato. B. Jowett, trans. 2 vols. New York: Random House.

Descartes, R.
    1967   The Philosophical Works of Descartes. E. S. Haldane and G. R. T. Ross, trans. 2 vols. Cambridge, Mass.: Cambridge University Press.

Esser, A. H.
    1974   Environment and mental health. Science, Medicine, and Man 1: 181–193.

Falck, B., and N. A. Hillarp.
  1959   On the cellular localization of catecholamines in the brain. Acta Anat. 38:
         277–279.

Ferrier, D.
  1876   The Functions of the Brain. London: Smith, Elder, and Company.

Foerster, H. von, P. M. Mora, and L. W. Amiot.
  1960   Doomsday: Friday, 13 November, A.D. 2026. Science 132: 1291–1295.

Freud, S.
  1949   New Introductory Lectures on Psychoanalysis. W. J. H. Sprott, trans.
         London: The Hogarth Press and The Institute of Psycho-Analysis.

  1953   The Interpretation of Dreams. (1900) Standard Edition. London: Hogarth
         Press.

Grange, K. M.
  1961   Pinel and eighteenth-century psychiatry. Bull. Hist. Med. 35: 442–453.

Greenbie, B.
  1976   Design for Diversity. Amsterdam-Oxford-New York: Elsevier Scientific
         Publishing Company.

Herrick, C. J.
  1933   The functions of the olfactory parts of the cerebral cortex. Proc. Nat. Acad.
         Sci. USA 19: 7–14.

Hinde, R. A.
  1972   Non-Verbal Communication. Cambridge, Mass.: The University Press.

Jeans, Sir James
  1943   Physics and Philosophy. Cambridge; The University Press.

Juorio, A. V., and M. Vogt.
  1967   Monoamines and their metabolites in the avian brain. J. Physiol. 189:
         489–518.

Lorenz, K. Z.
  1937   The companion in the bird's world. Auk 54: 245–273.

MacLean, P. D.
  1952   Some psychiatric implications of physiological studies on frontotemporal
         portion of limbic system (visceral brain). Electroenceph. Clin.
         Neurophysiol. 4: 407–418.

  1954   The limbic system and its hippocampal formation. Studies in animals and
         their possible application to man. J. Neurosurg. 11: 29–44.

  1958   Contrasting functions of limbic and neocortical systems of the brain and
         their relevance to psychophysiological aspects of medicine. Amer. J. Med.
         25: 611–626.

  1962   New findings relevant to the evolution of psychosexual functions of the
         brain. J. Nerv. Ment. Dis. 135: 289–301.

  1964   Mirror display in the squirrel monkey, Saimiri sciureus. Science 146:
         950–952.

  1967   The brain in relation to empathy and medical education. J. Nerv. Ment.
         Dis. 144: 374–382.

1970   The triune brain, emotion, and scientific bias. *In* The Neurosciences Second Study Program. F. O. Schmitt, ed. Pp. 336–349. New York: The Rockefeller University Press.

1972a   Cerebral evolution and emotional processes: new findings on the striatal complex. Ann. N.Y. Acad. Sci. 193: 137–149.

1972b   Implications of microelectrode findings on exteroceptive inputs to the limbic cortex. *In* Limbic System Mechanisms and Autonomic Function. C. H. Hockman, ed. Pp. 115–136. Springfield: Charles C Thomas.

1973a   Effects of pallidal lesions on species-typical display behavior of squirrel monkey. Fed. Proc. 32: 384.

1973b   The brain's generation gap: some human implications. Zygon J. Relig. Sci. 8: 113–127.

1973c   A triune concept of the brain and behaviour: Lecture I. Man's reptilian and limbic inheritance; Lecture II. Man's limbic brain and the psychoses; Lecture III. New trends in man's evolution. *In* The Hincks Memorial Lectures. T. Boag and D. Campbell, eds. Pp. 6–66. Toronto: University of Toronto Press.

1974   The triune brain. *In* Medical World News, Special Supplement on "Psychiatry" 1: 55–60.

1975   Role of pallidal projections in species-typical behavior of squirrel monkey. Trans. Amer. Neurol. Assoc. 100: 110–113.

1976   The imitative-creative interplay of our three mentalities. *In* Astride the Two Cultures: Arthur Koestler at 70. Harold Harris, ed. Pp. 187–213. New York: Random House.

Mallows, E. W. N.
1970   Urban planning and the systems approach. (I.B.M. System & Engineering Symposium, October, 1969) *In* Plan (Successor to S.A. Archit. Rec.) 55: 11–24.

Meadows, D. H., D. L. Meadows, J. Randers, and W. W. Behrens, III.
1972   The Limits to Growth. New York: Universe Books.

Monod, J.
1971   Chance and Necessity. New York: A. A. Knopf, Inc..

Morsbach, H.
1973   Aspects of nonverbal communication in Japan. J. Nerv. Ment. Dis. 157: 262–277.

Myers, K., C. S. Hale, R. Myktowycz, and R. L. Hughes.
1971   The effects of varying density and space on sociality and health in animals. *In* Behavior and Environment. A. H. Esser, ed. Pp. 148–187. New York: Plenum Press.

Parent, A. and A. Olivier.
1970   Comparative histochemical study of the corpus striatum. J. Hirnforsch. 12: 75–81.

Piaget, J.
1967   Six Psychological Studies. A. Tenzer, trans. New York: Random House.

Ploog, D. W., and P. D. MacLean.
    1963   Display of penile erection in squirrel monkey *(Saimiri sciureus)*. Anim.
            Behav. 11: 32–39.

Shakow, D., and D. Rapaport.
    1964   The influence of Freud on American psychology. *In* Psychological Issues.
            Vol. 4, monograph 13. New York: International Universities Press.

Snow, C. P.
    1967   Variety of Men. New York: Charles Scribner's Sons.

Spencer, H.
    1896   Principles of Psychology. 2 vols. New York: D. Appleton and Company.

Tinbergen, N.
    1951   The Study of Instinct. Oxford: The Clarendon Press.

Washburn, S. L.
    1973   The evolution game. J. Hum. Evol. 2: 557–561.

Washburn, S. L., and R. L. Ciochon.
    1974   Canine teeth: notes on controversies in the study of human evolution.

Watson, J. B.
    1924   Behaviorism. New York: The People's Institute Publishing Company.

Weisberger, B. A.
    1972   Book Review of "Black Mountain." The Washington Post, November 19.

Wiener, N.
    1948   Cybernetics, or Control and Communication in the Animal and the
            Machine. New York: Wiley.

Facial threat expression of rhesus monkey.

Ronald E. Myers

Comparative Neurology of Vocalization and Speech:
Proof of a Dichotomy

Rhesus monkeys and chimpanzees have excellent control over the use of their hands, but only meager volitional control over their vocal apparatus. Face and voice function primarily as signalling systems in social and emotional situations. Myers shows that in man these emotional uses of face and voice continue, but a new system of communication is added based on cortical mechanisms and voluntary control.

EMOTIONAL AND VOLITIONAL FACIAL EXPRESSION

The existence of a dichotomy between those behaviors and body movements that are volitional and those that are emotional has been recognized in clinical neurology for many years. The neurologist, in examining his patient, requests him to "show his teeth" with all his strength. In response to this command, the patient strongly bares his teeth. If a weakness appears in the action of one or both sides of the face, the neurologist concludes that the patient has a paralysis of volitional use of the face—if no evidence exists of a

59

lesion of the facial nerve or its nucleus. The neurologist also waits for occasions when the patient smiles or laughs in response to an amusing situation. Then he determines whether the patient shows any asymmetry in action of the two sides of his face. If such an asymmetry appears, and the patient does not also show evidence of a volitional paralysis, the neurologist concludes that the patient has a weakness in the emotional use of the opposite side of the face. Such a paralysis can be bilateral or entirely one-sided. These time-honored methods of investigating the motor activity of the face serve as the basis for an operational definition of the terms *volitional* and *emotional*. Those unintended facial movements (or vocal utterances) which arise as a part of an instinctual reaction to an appropriately evocative situation may be regarded as emotional or involuntary, while those facial movements (or vocal emissions) which are clearly intended by the individual or requested of him may be considered volitional.

The neurologist has long recognized that large lesions that affect the precentral gyrus, the internal capsule, or the corticospinal tracts of one or the other hemisphere may cause a paralysis of the voluntary use of the opposite side of the body. However, if such lesions are restricted to inferior portions of the precentral convolution or to a small part of the internal capsule, they may cause a paralysis restricted to the opposite side of the face. Lesions affecting the left hemisphere in such locations also often cause disorders of speech. The association of right facial weakness with speech difficulties is common and excites no special attention in the literature— particularly since such right-sided motor symptomatologies associated with speech disorders have been described since the early 1800s. See, for example, Broca's 1865 description of the occurrence of speech disorders following lesions of the posterior portion of the left inferior frontal convolution.[1]

It took neurologists a longer time to recognize that isolated paralyses of the emotional use of the face may occur. However, by the middle 1800s the possibility of such paralyses was well recognized, and from that period through the 1920s numerous papers describing this phenomenon appeared under a variety of titles.[2-4] In contrast to the ready recognition of a relation between paralyses of volitional movement and the corticospinal system, it has been difficult to define a precise neurologic basis for paralyses affecting emotional expression. This has been true partly because isolated paralyses of emotional expression are uncommon, and because they result from a diversity of pathologic processes. Also, the neurologic mechanisms which underlie emotional expression are widely distributed in the brain and often encompass structures that are small in size and that occur intermixed with diverse other brain tracts and nuclei. However, the mechanisms which underlie emotional expression are generally located in or near the midline, and they occupy ventral regions of the brain. Brain loci, lesions of which cause paralyses of emotional expression, include the orbitofrontal region of frontal lobes, the area innominata, the medial aspect of the tip of temporal lobes, the anterior ventral thalamus, the subthalamus and hypothalamus, and restricted regions of the brain stem. Thus, the clinical neurologist has clearly recognized and documented a dichotomy of function in the use of the face, and he has provided information as to the locations in the brain of mechanisms that underlie the two types of function (Figure 1).

VOCAL EXPRESSION

If a clear-cut dichotomy has been defined with respect to facial expressions, such a distinction remains less clear regarding functions of the vocal apparatus. In part, this lack of definition of a distinction between vocalizations results from the major attention that has been paid in man to aphasic difficulties. However, because of this preoccupation with aphasia, we now

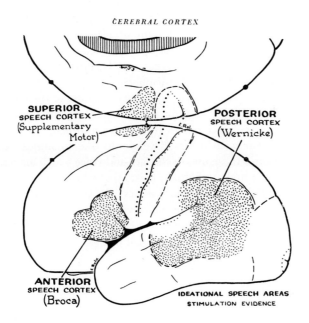

FIGURE 1
Left hemisphere schema of man illustrating those areas of cortex which, when electrically stimulated, interfere with ongoing speech (from Penfield and Roberts).

know a great deal about the cerebral mechanisms that underlie the use of the human voice for symbolic communication. Several specific cortex regions of the left hemisphere have been positively identified as supporting speech as illustrated in Figure 1. A detailed discussion of these regions and their functions are available in Penfield and Roberts[5] and Nielsen.[6]

One branch of the dichotomy in the use of the vocal apparatus in man is symbolic communication, i.e., speech. The use of words in verbal communications is clearly volitional. The existence of a second type of use of the voice, i.e., in emotional expression, remains uncertain and its neurology poorly defined. Indeed, the neurologist, when confronted with the proposition of an emotional use of the voice, inevitably thinks of the use of curse words or interjections. Such an interpretation is particularly reasonable since the use of curse words may be preserved in aphasics who have otherwise lost all evidence of speech. Still, it must be remembered that curse words are symbolic utterances denoting specific meanings, which in their nonpathologic use are tied in with volitional communication. Thus, from a biologic view, the other branch of the dichotomy in the use of the human vocal apparatus is probably restricted to such phenomena as laughing, screaming, shouting, and other nonsymbolic utterances that may arise involuntarily during the expression of mirth or joy in happy situations, or of fear, surprise, or anger in less pleasant situations. To my knowledge, no publications are available that bear upon the neurology of such emotional uses of the voice except insofar as they may become exaggerated following diffuse bilateral lesions of the hemispheres.[7]

What are the relations existing between the use of the face and voice? The almost total absence of information regarding the neurology of the emotional use of the voice makes this question difficult to answer. However, in agreement with others,[8] we would like to suggest that strong functional ties bind those uses of the face and voice which are emotional in character. Thus, a high probability exists that when the face is used to express emotion, the voice also participates. On the other hand, such close relations between face and voice use in volitional acts are not so obvious. The close links which bind emotional face and voice use lead us to suggest that the neurology underlying emotional face use is similar to that which controls the emotional use of the voice.

HUMAN AND NONHUMAN RESPONSES

The presence of two distinct uses of the face in man and of a clear separateness in the underlying neural mechanisms raises questions about the significance of facial expressions and vocalizations in animals and of their relations to facial expressions and speech in man. Are the vocalizations and facial expressions of animals identified with the emotional or with the volitional use of the voice and face in man? Or, are the vocal responses of animals still organized differently from either of the uses of the voice in the human?

Several theories have been proposed to relate the vocalizations of animals and speech in man. The most obvious of these suggests that animal vocal responses gradually evolved into the speech of man. Hockett has advanced a view that suggests such an interpretation.[9] Another interpretation proposes that human speech originated not from the vocalizations of animals but from gestures and other body movements.[10] A third interpretation views human speech as having evolved *de novo* and as bearing little relation to animal vocal responses. This interpretation has been advocated most recently by Lancaster[11] and by Myers.[12] Though hypotheses supporting the *de novo* origin of speech have not enjoyed widespread acceptance, the data which support such an evolution are considerable. The remainder of the present discussion reviews neurologic evidence bearing upon this question.

A dichotomy exists in the use of the face and voice in man. Both the face and the voice may be activated by two separate brain mechanisms, one related to volition and the other related to emotion. What is the situation of the monkey with respect to its use of face and voice? It is difficult to draw inferences from animal observation with respect to the nature of their actions vis-à-vis volition versus instinct. It is possible only to suggest that the use of both the face and the voice by rhesus monkeys in their natural habitat seems to be restricted to circumstances that connote emotion. Their facial expressions and vocalizations typically appear as components of behavior patterns which are best characterized as emotional or instinctual in nature. For example, this species uses its facial expressions and vocalizations as component parts of attack reactions, of defense reactions, of fear reactions, in the expression of loneliness or isolation from the group, or of hunger. Figure 2 shows an example of the rhesus monkey's use of facial expression as a component part of a reaction pattern expressive of hostility or aggression. At the same time, the rhesus monkey does not seem to utilize its facial expressions or vocalizations outside of such patterns of emotional reaction. Therefore, observations of rhesus monkeys in a natural setting encourage the view that this species uses its face and voice almost exclusively as component parts of instinctual patterns of reaction.

The transfer of the rhesus monkey to experimental situations in the laboratory permits more direct inferences to be drawn regarding the presence or absence of a volitional use of its face and voice. The rhesus monkey (and other animals) may be taught to carry out a variety of movements to receive food reward or to avoid punishment. The rhesus monkey easily learns to press a lever with its hands and, within a short time, this response is performed with speed and accuracy. We believe this use of the hands in bar pressing is volitional, and the ease and rapidity of learning this

FIGURE 2

Facial threat expression of rhesus monkey. Rhesus monkeys use facial expressions and vocalizations almost exclusively to communicate socially and to express emotion. Photograph by Dr. John Vandenbergh, Dorothea Dix Hospital, Raleigh, North Carolina.

response indicates that the volitional mechanisms of the brain of the monkey have a ready access to those brain centers that organize hand movements. Can the rhesus monkey similarly be taught to utilize its facial expressions or its vocalizations mediately to achieve food reward or to avoid punishment? Unfortunately, to the knowledge of the writer, no such studies have been carried out with respect to utilization of facial expressions. However, studies on vocalizations have been accomplished and the results are conflicting.

Skinner was the first to suggest that marked difficulties may arise when attempts are made to condition the vocal responses of animals.[13] He explained this refractoriness by citing the fact that animal vocalizations are tied in with emotion or instinct and are not volitional. He states, "Well-defined emotional and other innate responses comprise reflex systems which are difficult, if not impossible, to modify by operant reinforcement. Vocal behavior below the human level is especially refractory."[13]

The results of work carried out by Shirley Myers and co-workers[14] appear to conflict with this point of view. These workers seem to have succeeded in establishing definite instrumental vocal responses in cebus monkeys. But careful study of this work indicates that the cebus monkeys' acquisition of these responses proceeded with difficulty, and only three of the original six animals studied succeeded in showing evidence of such a response. Yamaguchi and Myers, however, failed to establish discriminative vocal responses in six rhesus monkeys despite prolonged efforts at conditioning utilizing several different experimental approaches.[15] Figure 3 compares the ease with which differential bar pressing responses were developed depending on the presence or absence of a red light, and the difficulties that were encountered in establishing comparable differential responses using their vocalizations. These latter authors emphasize the need for the animals' responses to be discriminative in nature in order to distinguish more definite conditioning of vocal responses from enhancements of vocalizations that may develop in association with acquired anticipatory sets. That is, the animals may come to expect to be fed or to receive electric shocks when they are placed in a specific setting at a specific time. Such

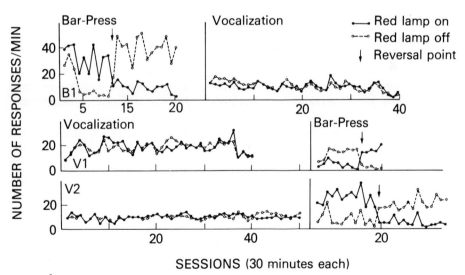

COMPARISONS OF BAR-PRESS WITH VOCALIZATION

FIGURE 3

Performance of three rhesus monkeys on differential bar-press and vocalization conditioning. At the arrows, the reward-punishment values of the two conditions were reversed. Differential responses according to whether a red lamp was on or off easily emerged in relation to bar-press behavior but not to vocalization. (From Yamaguchi and Myers[15])

expectations induce certain tensions or emotional states in the animals. These states then may lead to enhanced facial expressions or vocal responses. Thus, increases or decreases in number or prominence of vocal responses in different situations need not be adjudged to be examples of the use of vocalizations as instrumental responses but, rather, they may indicate only enhanced or aroused feelings in the animals.

Sutton et al. have also attempted to condition the vocal responses of rhesus monkeys.[16] They described what they interpreted to be a definite success. However, they used an experimental paradigm which did not force a clear-cut discriminative aspect to the emission of the vocal responses and, from the set-up of the experiment, it is also possible to interpret the resulting enhanced vocal responses as an anticipatory set. Further, they used only three animals in their study and only one behaved in a way that could possibly be interpreted as suggestive of conditioning.

A further study describes the results obtained using 15 rhesus monkeys in a conditioning situation where individual vocal responses were food rewarded.[17] These animals were studied during from 17 to 37 thirty-minute sessions. Under these conditions, 4 of the 15 animals increased the numbers of vocal responses which they emitted over the first 4–7 sessions of testing. This early response enhancement suggested that these animals had, indeed, acquired conditioned vocal responses. However, after this brief initial period, all of the animals stabilized at new rates of responding. The slowness of these new rates would have easily permitted these animals to further augment the frequencies of their responses should they have been so inclined. The remaining 11 animals failed to show any definite enhancements of their vocal responses over the numerous sessions of testing. Despite food reward, one of the animals progressively decreased the numbers of vocalizations it emitted during the sessions of testing. Figure 4 presents the vocalization curves of two animals, one of which showed an incremental pattern of responding and the other a decremental pattern of responding during testing. These overall results indicate that if rhesus monkey vocal responses are available for instrumental use and can be brought under volitional control, they are poorly so. Our overall experience with vocal conditioning in rhesus monkeys supports the view that the vocal apparatus (and in all likelihood, the facial musculature as well) of nonhuman primates is poorly accessible to those mechanisms of the brain that organize and control voluntary movements.

BRAIN MECHANISMS

Until recently little information has been available to identify those mechanisms of the brain that control the vocal responses of nonhuman primates. The earliest data bearing on this topic came from the brain

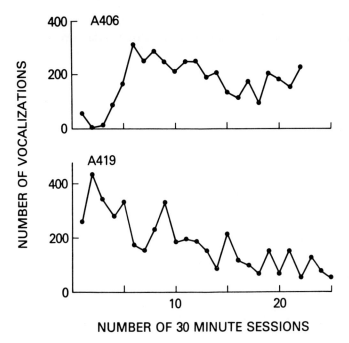

FIGURE 4

Two patterns of rhesus monkey vocalization which emerged in association with food reward. In the incremental pattern (above), the vocal responses progressively augmented over the early sessions of testing and then stabilized at intermediate rates of responding. In the decremental pattern (below), the numbers of vocal responses emitted progressively declined despite food reinforcement. The incremental pattern occurred in 4 and the decremental in 1 of 15 monkeys tested. The remaining 10 animals showed essentially no change in numbers of vocalizations from beginning to end of testing despite food reward.

stimulation studies of Robinson using rhesus monkeys[18] and of Jurgens et al. using squirrel monkeys.[19,20] The results of these studies show that it is the stimulation of white matter tracts and subcortical grey matter structures related to the limbic system that produce vocalizations. Robinson also indicates that direct stimulation of corticospinal fibers at the level of the internal capsule or of the pes pedunculi fails to elicit vocal responses. Jurgens[21] and Walker and Green[22] also failed to produce vocalizations when the lower precentral gyrus (the motor face area) was stimulated in monkeys, although movements of the vocal folds have been observed by others.[23,24] It is generally held that the precentral gyrus and the corticospinal tracts are intimately tied up with voluntary movements while the various structures of the limbic system regulate instinctive behavior. If these suppositions are true, then these stimulation studies suggest that the vocalizations of rhesus monkeys are controlled by brain structures having to do with instinctual or emotional behavior. They afford no positive evidence of a linkage with neural mechanisms having to do with volition.

What contributions have ablation studies made toward elucidating the brain mechanisms which underlie animal vocalizations? Few studies have been carried out in this area utilizing nonhuman primates. Nonetheless, studies with cats have demonstrated the presence of mechanisms in the midbrain—located particularly in the periaqueductal grey substance—which are involved in organizing facio-vocal activity.[25-27] These results remind us that humans also may develop paralysis of the emotional use of the face following lesions at the level of the midbrain or pons.[2,28,29]

Regions can be identified over the hemispheral surfaces of monkeys which correspond in their general positioning to the cortical speech areas of man. These regions include the inferior third of the precentral gyrus, the supplementary motor face area located on the medial surface of the hemisphere, the supraangular, and the supramarginal regions. Myers,[11] and Yamaguchi and Myers[30] have described studies in which these individual monkey "speech" areas have been removed bilaterally, alone, and in various combinations. These studies failed to identify any clearly defined deficits in vocal responses following lesions of any single one of these zones of cortex. On the other hand, as one after the other of these speech areas were removed in sequence in the same animal, the vocal responses did decline in number. The depressions following removal of all four areas averaged 49% and ranged between 18–77% of the initial rates. These depressions might suggest some specific relation of these areas to vocalizations, but they also may be part of a generalized depression of neurologic function resulting from such extensive bilateral cortical removals. Sutton, Larson, and Lindeman also were unable to demonstrate any significant deficits in vocal responses following speech area lesions in monkeys.[31]

The motor homunculus of Penfield and Jasper[32] and the corresponding simunculus of Woolsey et al.[33] contribute to our knowledge of the cortical mechanisms underlying movement. The boundaries of the projections of the homunculi and simunculi upon the cortical surface have been carefully defined from electrical stimulation carried out in many individuals. It might be anticipated from the conformation of the simunculus that lesions of the lower one-third of the precentral gyrus of monkeys should cause a weakness of the opposite side of the face. Such a result would correspond with the findings observed in human beings who sustain pathologic lesions in this region. However, specific surgical topectomies affecting this zone and, indeed, ablations of the entire precentral gyrus in rhesus monkeys, fail to produce such striking contralateral facial weaknesses.[13,34] Rather, deficits in contralateral facial motor activity were observed only in animals

following removal of the temporal lobe, of the entire frontal lobe, or, surprisingly, of the prefrontal-orbitofrontal cortex alone. The lack of definite
facial weakness following precentral gyrus removal in the monkey is illustrated in Figure 5. These findings suggest that the rhesus monkey controls
its facial movements not through precentral gyrus mechanisms having to
do with volitional movement, but through mechanisms in areas of the
brain which regulate instinctual and emotional behavior. The identification of the involvement of the orbitofrontal, the anterior temporal, and the
cingulate cortex in the control of instinctual and emotional behavior is described elsewhere.[34-38]

The effects of lesions in the inferior precentral gyrus (the cortical area for
volitional control of face in man) on movement of the opposite side of the
face and of the speech areas on vocalizations of rhesus monkeys are unim-

FIGURE 5
Facial expression of juvenile rhesus monkey following removal of right precentral gyrus eight
days earlier. Note the absence of any asymmetry of expression.

pressive. This, coupled with the multiple suggestions that monkey facial expressions and vocalizations are regulated by brain centers concerned with social behavior and emotion, have led to a more specific examination of the relations between those neural mechanisms which control emotion and the emission of vocal responses by rhesus monkeys. These studies were carried out with animals that were vocalizing at stable and predictable rates within a standard test situation. Zones of cortex which have been earlier identified as controlling social behavior and emotion were then removed (Figure 6). In contrast to the mildly depressive effects of removals of the speech areas as

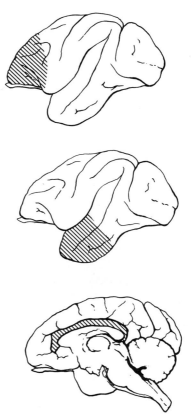

FIGURE 6
In monkeys, lesions of prefrontal-orbitofrontal cortex
bilaterally (above) markedly and permanently reduced all
facets of social-emotional behavior and largely abolished
vocalizations and facial expressions. Lesions of anterior
temporal cortex (middle) produced similar but less marked
deficits while cingulate cortex ablations (below) exerted a
variable effect on vocalizations.

described above, the lesions in zones of cortex controlling social behavior and emotion produced marked and often permanent deficits in vocalizations.[17] Of the three areas of cortex that regulate social behavior, it was the lesions of the prefrontal-orbitofrontal region* that produced the most marked deficits. In many instances, the effects of this lesion on spontaneous vocalizations were devastating enough to cause near total and permanent muteness (Figure 7). The removals of anterior temporal or cingulate cortex produced less marked temporary significant deficits. The declines in vocalization were observed not only in the formal testing situation in the psychologic laboratory but they also showed up in studies of the social behavior and emotion of monkeys maintained in the free-ranging situation[35,36] or held in social groups in enclosures.[37] Furthermore, not only did these lesions produce major deficits in the vocalizations of these animals but they also significantly diminished their use of facial expressions as tested in appropriately evocative situations.

CHIMPANZEE STUDIES

What do studies using nonhuman primate species other than rhesus monkeys contribute to our knowledge regarding the volitional use of the face and voice? Of major interest have been the recent studies on the chimpan-

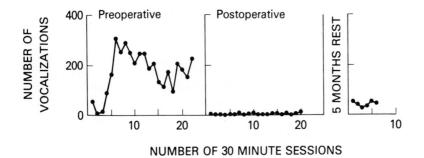

FIGURE 7
Effects of removal of prefrontal-orbitofrontal cortex on vocalizations in monkeys. The losses of vocal responses were sometimes nearly total and seemingly permanent as illustrated above. It is the removal of the orbitofrontal cortex that largely accounts for the social behavioral and vocalization deficits that are seen following removal of the entire prefrontal lobe.

---

*It is removal of the orbitofrontal component of the prefrontal-orbitofrontal region that accounts for the deficits in social behavior and vocalization which follow lesions of the prefrontal-orbitofrontal region as has been demonstrated by further work in this laboratory.[39]

zee's use and nonuse of hand gestures and vocal utterances for interspecific communication. The Hayeses were the earliest to rear a chimpanzee in a familial setting and to attempt exhaustively to teach it to use its voice to communicate.[40] The Hayeses generally failed to teach their chimpanzee to use word sounds although they did report that the chimpanzee had expended considerable effort (without success) to teach them (the Hayeses) to communicate with it (the chimpanzee) not through imposed word forms but through hand gestures.[41]

Others have taken up this pursuit.[42-44] What has generally emerged from these studies is that the chimpanzee is a highly intelligent animal that can communicate ideas very well with humans provided it is not required to use its vocal apparatus. The chimpanzee has shown a particular cleverness in using its hands to communicate both by means of sign language and by depressing levers. These overall results fit well into the schema evolved in this paper. That is, the chimpanzee, like the monkey, tells us: "If you want me to communicate with you, don't ask me to do so using my vocal apparatus and word forms but rather ask me to communicate using any system you may want which requires the manipulation of levers or the use of hand gestures." The chimpanzee confirms that it, like the rhesus monkey, has excellent volitional control over its hands and body movements but only meager volitional control over its vocal apparatus. This view reminds us of the ease with which the rhesus monkey can be taught to bar press for food reward and of the enormous difficulty encountered when attempts are made to condition it to utilize its vocal apparatus according to a discriminative schedule in instrumental conditioning.

CONCLUSIONS

In conclusion, the face and voice of nonhuman primates are under poor volitional control. In these species the face and voice function primarily as signaling systems but apparently only within the realm of social and emotional behavior. Both brain stimulation and lesion studies support this view. The facial expressions and vocalizations of nonhuman primates are closely akin to man's emotional use of the face and voice. It remains unclear whether or at what level the nonhuman primate has developed even rudimentary mechanisms in its brain that can support any voluntary control of its face or voice. Certainly, if further work demonstrates the existence of such mechanisms, it must be concluded from the present review that the magnitude of their influence is small.

What do the present considerations suggest regarding the origins of speech in man? If the present interpretations are valid, it is apparent that the speech of man has not evolved from the vocal responses of lower primates. Rather, speech has developed *de novo* in man during his evolutionary development beyond the level of monkeys or, apparently, of the apes. From a neurologic view, the evolution of speech represents the evolution of those mechanisms of the cerebrum located posteriorly in zones of cortex that function to analyze the information of the senses, to establish memories thereof, and to organize voluntary responses proceeding from these analyses or memories. The separate and distinct mechanisms of the cerebrum that control emotional and instinctive behavior still remain in humans to link us phylogenetically with our lower primate forebears. However, for the proper study of language, one would do well to focus on the physiologic properties of the posterior cerebrum of man himself and to view with suspicion speculations pointing out similarities between vocal responses of animals and speech in man.

References Cited

1. Broca, P. 1865. Sur la faculte du langage. Bull. Soc. Anthropol. 4: 493–494.

2. Spiller, W. G. 1912. Loss of emotional movements of the face with preservation or slight impairment of voluntary movement in partial paralysis of the facial nerve. Am. J. Med. Sci. CXLIII: 390–393.

3. Monrad-Krohn, G. H. 1924. On the dissociation of voluntary and emotional innervation in facial paresis of central origin. Brain 47: 22–35.

4. Feiling, A. 1927. A case of mimic facial paralysis. J. Neurol. & Psychopath. 8: 141–145.

5. Penfield, W., and L. Roberts 1959. Speech and Brain Mechanisms. Pp. 119–137. Princeton, N.J.: Princeton University Press.

6. Nielsen, J. M. 1946. Agnosia, Apraxia, Aphasia: Their Value in Cerebral Localization. Second ed. Pp. 1–201. New York: Paul H. Hoeber.

7. Wilson, S. A. K. 1929. Pathological Laughter and Crying. *In* Modern Problems in Neurology. Pp. 260–296. New York: Williams Wood and Co.

8. Magoun, H. W., D. Atlas, E. H. Ingersall, and S. W. Ranson. 1937. Associated facial, vocal, and respiratory components of emotional expression: an experimental study. J. Neurol. Psychopath. 17: 241–255.

9. Hockett, C. F. 1960. The origin of speech. Sci. Am. 203:88–96.

10. Hewes, G. W. 1973. Primate communication and the gestural origin of language. Current Anthropology 14:5–24.

11. Lancaster, J. B. 1968. Primate communication systems and the emergence of human language. *In* Primates. P. C. Jay, ed. Pp. 439–457. New York: Holt.

12. Myers, R. E. 1969. Neurology of social communication in primates. Proc. 2nd Int. Congr. Primat. 3:1–9. Basel: Karger.

13. Skinner, B. F. 1957. Verbal Behavior. P. 463. New York: Appleton-Century-Crofts.

14. Myers, S. A., J. A. Horei, and H. S. Pennybacker 1965. Operant control of vocal behavior in the monkey Cebus albifrons. Psychonomic Sci. 3:389–390.

15. Yamaguchi, S., and R. E. Myers 1972. Failure of discriminative vocal conditioning in rhesus monkey. Brain Res. 37:109–114.

16. Sutton, D., C. Larson, E. M. Taylor, and R. C. Lindeman 1973. Vocalization in rhesus monkey: Conditionability. Brain Res. 52:225–231.

17. Yamaguchi, S., and R. E. Myers. In press. Cortical mechanisms underlying vocalization in rhesus monkey: prefrontal-orbitofrontal, anterior temporal, and cingulate cortex. Neuropsychologia.

18. Robinson, B. 1967. Vocalization evoked from forebrain in Macaca mulatta. Physiol. Behav. 2:345–354.

19. Jurgens, U., M. Maurus, D. Pioog, and P. Winter 1967. Vocalization in the squirrel monkey (Saimiri sciureus) elicited by brain stimulation. Exp. Brain Res. 4:114–117.

20. Jurgens, U., and D. Ploog 1970. Cerebral representation of vocalization in the squirrel monkey. Exp. Brain Res. 10:426–434.

21. Jurgens, U. 1974. On the elicitability of vocalization from the cortical larynx area. Brain Res. 81:564–566.

22. Walker, A. E., and H. D. Green 1938. Electrical excitability of the motor face area: A comparative study in primates. J. Neurophysiol. 1:152–165.

23. Hast, M. H., and B. Milojevic 1966. The response of the vocal folds to electrical stimulation of the inferior frontal cortex of the squirrel monkey. Acta Oto-Laryng. (Stockholm) 61:196–204.

24. Hast, M. H., J. M. Fischer, A. B. Wetzel, and V. E. Thompson 1974. Cortical motor representation of the laryngeal muscles in Macaca mulatta. Brain Res. 73:229–240.

25. Kelly, A. H., L. E. Beaton, and H. W. Magoun 1946. A midbrain mechanism for facio-vocal activity. J. Neurophysiol. 9:181–189.

26. Kanai, T., and S. C. Wang 1962. Localization of the central vocalization mechanism in the brain stem of the cat. Exp. Neurol. 6:426–434.

27. Skultety, F. M. 1963. Stimulation of periaqueductal gray and hypothalamus. Arch. Neurol. 8:608–620.

28. Mills, C. K. 1912. Preliminary note on a new symptom complex due to a lesion of the cerebellum and cerebello-rubro-thalamic system, the main symptoms being ataxia of the upper and lower extremities of one side, and on the other side deafness, paralysis of emotional expression in the face and loss of the senses of pain heat and cold over the entire half of the body. J. Nerv. & Ment. Dix. XXXIX: 73–76.

29. Wilson, S. A. K. 1924. Pathologic laughing and crying. J. Neurol. & Psychopath. IV: 299–333.

30. Yamaguchi, S., and R. E. Myers 1975. Effects of "speech" area lesions on vocalization in monkey. Brain Res.

31. Sutton, D., C. Larson, and R. C. Lindeman 1974. Neocortical and limbic lesion effects on primate phonation. Brain Res. 71:61–75.

32. Penfield, W., and H. Jasper 1954. Epilepsy and the Functional Anatomy of the Human Brain. Pp. 52–106. Boston: Little, Brown & Co.

33. Woolsey, C. N., P. H. Settlage, D. R. Meyer, W. Spencer, T. Pinto-Hamuy, and A. M. Travis 1952. Patterns of localization in precentral and "supplementary" motor areas and their relation to the concept of a premotor area. Research Publications, Assoc. for Research in Nervous and Mental Diseases 30:238–264.

34. Myers, R. E. 1972. Role of prefrontal and anterior temporal cortex in social behavior and affect in monkeys. Acta Neurobiol. Exp. 32:567–579.

35. Myers, R. E., and C. S. Swett, Jr. 1970. Social behavior deficits of free-ranging monkeys after anterior temporal cortex removal. A preliminary report. Brain Res. 18:551–556.

36. Myers, R. E., C. Swett, and M. Miller 1973. Loss of social group affinity following prefrontal lesions in free-ranging macaques. Brain Res. 64:257–269.

37. Franzen, E., and R. E. Myers 1973. Neural control of social behavior: prefrontal and anterior temporal cortex. Neuropsychologia. 11:141–157.

38. Myers, R. E. 1975. Neurology of social behavior and affect in primates: a study of prefrontal and anterior temporal cortex. In Cerebral Localization. K. J. Zulch, O. Creutzfeldt, G.C. Gailbraith, eds. Pp. 161–170. New York: Springer Verlag.

39. Ervin, F. R., M. Raleigh, H. D. Steklis 1975. The orbital frontal center and monkey social behavior. 5th Annual Meeting, Soc. Neurosci., New York, 1975. Pp. 554.

40. Hayes, K. J. 1950. Vocalization and speech in chimpanzees. Am. Psychol. 5: 275–276.

41. Hayes, G. 1968. Spoken and gestural language learning in chimpanzees. Paper read at meeting of Psychomonic Society, November.

42. Gardner, R. A., and B. Gardner. 1969. Teaching sign language to a chimpanzee. Science. 165:664–672.

43. Premack, D. 1971. Language in chimpanzee? Science. 172:808–812.

44. Rumbaugh, D. M. 1975. Language and the acquisition of language-type skills by a chimpanzee. Origins and Evolution of Language and Speech. S. R. Harnad, H. D. Steklis, and J. Lancaster, eds. Annals New York Acad. Sci., Vol. 28.

Dr. Foster with an informant. Tzintzuntzan, Michoacan, Mexico.

Mary LeCron Foster

The Symbolic Structure of Primordial Language

This chapter is written primarily for linguists. It may be divided into two parts. The first pages outline problems and methods of linguistic reconstruction and give the data for a new view of language relationships. The general reader will want to examine only these first pages and the conclusions, best shown in Table 7. Foster shows that far more reconstruction is possible than has usually been thought to be the case. This fits with the idea that all languages are more similar than is generally supposed, and that they are the communication systems of Homo sapiens. *Further, Foster's linguistic reconstruction shows ". . . an elaborate capacity of the paleolithic mind to organize and classify experience . . ." (p. 117). In spite of the fact that only readers with linguistic experience will be able to follow the analysis, we have included the technical portion because of its great potential importance in interpreting human evolution and because this evidence is not available elsewhere. Foster's exposition makes clear that the enormous acceleration of cultural invention paralleling the rise of* Homo sapiens *is not likely to have occurred without speech. The success of this reconstruction shows that contemporary language had a common origin at not too distant a time.*

INTRODUCTION

The anthropological construct of nature versus culture has tended to obscure the fact that the emergence of man as a symbolizing animal, responding to nature through the filter of his cognitively systematic organization of experience, was a psychobiological event, different in ultimate impact but not in kind from any other evolutionary development. Nascent culture required a genetically unique cognitive apparatus geared to permit expansion of the principles of experiential systematization that already existed in a germinal state. This new cognitive experience, symbolization, depended upon maximum utilization of the twin cognitive functions of analogy and opposition, found in increasing measure as mammalian evolution progressed, but now reaching a culminative sophistication in culture, and especially language. It may be assumed that the intensification of cooperative social experience in hunting brought about a deepening of consciousness of differences between *them* and *us* as well as of the similarities upon which group identity was founded.

That early language, the primary symbolic structure, depended upon this cognitive duality emerges clearly from reconstructed primordial linguistic forms. The reconstruction presented here utilizes the methodology of comparative linguistics, developed during the 19th century in Indo-European studies, combined with the structural insights of 20th-century linguistics, in which relationships between parts are studied as a complex of systematic patterns, subliminal for language users but yielding their secrets to scientific analysis.

Early linguistic symbols (phonemes), apparently parental to all present-day languages, are reconstructed from a group of languages whose genetic relationship to one another is extremely remote. The reconstructed symbols are found to be nonarbitrary. Their motivation depends upon a gestural iconicity between manner of articulation and a movement or positioning in space which the symbol represents. Thus, the hypothesis presented here implies that early language was not *naming* in the conventional sense but representation of one kind of physical activity by means of another, displaced in time but similar in spatial relationship.[1]

I have called these sound-meaning units *phememes,* adapting a little-used linguistic term for the smallest lexical and grammatical unit (Pei and Gaynor 1969:166) to the purpose. The primordial phememes are minimal units uniting distinctive features of sound and meaning. There are 18 phememes in three series of articulation type and six of articulation placement, as shown in Table 1.

The semantic and formal aspects of the phememe unite through their features which are taxonomically associated from the standpoint of the point of articulation. A more graphic table would represent the phememic set in the form of a cone, uniting labial and labial-velar series through their common peripherality. The excurvative feature of the labio-velars is expressed through mouth-rounding, while the contiguous feature of the labials is expressed by means of lip protrusion for *p, lip or tooth overhang for *f, and the bilateral positioning of the lips with simultaneous (focal) voicing in articulation of *m. While the approximatives (fricatives) had voiced, vocalic allophones, their consonantal articulation was probably voiceless, like that of the partitives (stops). The descriptive term associated in the table with each phememe does not represent a feature but a descriptive

TABLE 1

Primordial Phememes

	Proximal				Distal	
	Associative			Extensive	Protractive	Dissociative
	Peripheral	Internal				Peripheral
	Contiguous (Labial)	Intrusive (Dental)	Obtrusive (Alveolar)	Rectilineal (Palatal)	Disjunctive (Velar)	Excurvative (Labio-Velar)
(Partitive) (Stopped)	p — Projective (⊓)	t — Intromissive (↓ in circle)	c — Emissive (↑ in circle)	č — Clinal (/)	k — Delative (→)	kʷ — Circumversive (↻)
Approximative (Fricative)	f — Abessive (⊢×)	θ — Juxtapositive (→⊣)	s — Expansive (✛)	š — Vertical (|)	x — Oppositive (←→)	xʷ — Apertive (○)
Focal (Resonant)	m — Bilateral (‖)	n — Interpositive (⊠)	l — Extenuative (↕ dashed)	y — Horizontal (—)	r — Sublative (←)	w — Circumscriptive (●)

79

formula, or condensation, of the feature bundle into which that phememe may be analyzed.

The phememic semantics began to become apparent as the reconstruction progressed. Sound and meaning were isolated by means of commutation: the discovery procedure utilized in analysis of any unwritten language. Once the meaning of individual units had been more or less firmly established, the dimensions of the system of classification began to emerge. Individual phememes contrast with one another in five ways that are identifiable on both articulatory and semantic levels. I have attempted to select a feature terminology to define these contrasts which will bring out the iconicity of the two levels. I have also included standardized linguistic terminology in parentheses below the appropriate iconic term for that feature. Spatial (articulatory rather than auditory) designations for other features have been adopted in departure from the usual terminological practise in current linguistic feature analysis. Features define both mode of oral production of the sound and the activity or state that is to be symbolized. A graphic, spatial portrait is provided in the table for the spatial relationship defined through each phememe.

It was long believed that language must have a history of several million years, the span of time in which man has been a tool user (Sapir 1921:23; De Laguna 1963:49). Recent studies of primate behavior show that primates other than man, especially chimpanzees, are capable of learning the construction and use of simple tools by observation and without language (Lancaster 1968). The enormous acceleration of cultural invention that typified the emergence of *Homo sapiens* in the late Pleistocene could neither have occurred without speech nor without some major catalyst. If we put the two together we have the emergence of language proper, not as utterance or signal, which must have preceded it, but as system or symbol. This hypothesis reduces the history of language to a period of approximately fifty thousand years; a time span with a far greater probability of reconstructive accessibility than the several million years postulated earlier.

It seems likely that just as tool construction was very gradually achieving greater complexity, so oral signalling was improving to incorporate structural characteristics that ultimately underlay true language. A careful study of the cognitive principles necessary to the invention of successive tool types will undoubtedly show them to be very similar, perhaps identical, to the principles underlying early language, and it can be assumed that they existed in germinal form in prelanguage.

Reconstruction of PL (primordial language) requires an assumption of monogenesis. If all extant languages are assumed to be related because descended from a common ancestor, linguistic forms from a random group of languages may be assembled as the data upon which the reconstruction is to be based, provided that the members do not share a known history. The more divergent the languages, both genetically and geographically, the more plausible the hypothesis that what is reconstructed is truly PL rather than a more recent stage of language.

A great deal of trial and error has gone into the PL reconstruction presented here. The frequent thrill of discovery as successive bits of the puzzle began to slip more fruitfully and surely into position has served to counteract the loneliness of the enterprise; for reconstruction of remote history does not quite have linguistic respectability. The received tradition provides that language history be successively unveiled backward in time through a long series of reconstructed stages. The assumption is that as past stages are recovered, they are to be compared to one another: a laborious process that comparative linguists have been hesitant to bypass.

My reluctance to counter tradition would probably have been as great as that of any other linguist but for the accidental discovery that resemblances between presumably unrelated American Indian languages were too wide-spread to be due to chance borrowing. I had begun to mass-compare vo-cabularies in order to trace the type and direction of lexical loans in North America, a spare time activity that quickly provided an easily scanned assemblage of phonological reflexes for many basic semantic glosses in a great variety of languages. It seemed to me, a convinced comparativist, that if the striking similarities that I found were unexplainable by borrowing, they could only be attributed to genetic relationship and it should, there-fore, be possible to recover the ancestral forms by rigorous application of the comparative method.

I soon found that the phonological similarities between words, and especially verbal bases (stems or roots) of similar meaning, were not limited to the American continent but had a worldwide extension. Quite unexpec-tedly I found myself committed to the theory of monogenesis and became bent upon discovery of the primordial system from which modern forms might plausibly and regularly derive, for the cornerstone of the 19th-century comparative method is regularity of sound change. The significant contribution of the 20th century to this solid scientific base is the notion that the regularities of change imply systematization at any given syn-chronic stage, even though the system is always imperfect in some respects.

I must assume, then, a primeval phonological system with an inner logic of simplicity and symmetricality. Phonological complexity at the outset of language seemed highly unlikely. Throughout my reconstructive efforts, then, I worked to discover ways of reducing rather than augmenting the phoneme inventory of primordial language (PL). The only way that this can be done is to assume that new phonemes in particular languages arose from morphophonemic coalescence of phoneme clusters. This hypothesis has stood me in good stead and my pot of gold at the end of the primeval rainbow has been the unexpected discovery of the logic of the semantic system to which the phonological system is analogically connected.

Acceptance of the theory presented here entails rejection or revision of several cherished linguistic truisms: (1) that the earliest viable stage of language is unrecoverable; (2) that the major reason for this is that for each earlier reconstructed stage fewer morphemes are available because of erosion and replacement; and (3) that a meaningful segment of language is sign rather than symbol because the sounds by which it is conveyed are arbitrarily assigned. Equally questionable assumptions with wide, if not universal, acceptance are: (4) that the paradigmatic aspect of language is less interesting than its syntax; and (5) that the Anatolian languages, including Hittite, constitute an Indo-European subdivision much as any other, albeit more archaic (cf. Puhvel 1966).

The dictum of the impossibility of recovery of early language has discouraged systematic application of the comparative method of languages assumed to be unrelated. Thus, although Swadesh (1971) was convinced of monogenesis, he discounted the comparative method as an effectual reconstructive tool at such a great time depth. The widespread assumption that phonological and semantic erosion have been so extensive as to make reconstruction of very early language strata impossible is made despite strong contrary evidence. The contravening data are from the only reasonably completely reconstructed language of significant time depth: PIE (Proto-Indo-European), assumed to have been spoken during the third millenium B.C. A virtually complete stock of reconstructed morphemes not only exists for intermediate stages such as Germanic or Romance, but is no less ample for the parent language itself. With erosion, one would expect at least some diminution in recoverable morpheme stock from the later to the earlier periods. Yet this is not the case, despite known lexical borrowing and obsolescence.

The explanation for this seeming anomaly lies in the fact that reconstruction deals largely with morphemes—minimal meaningful partials—especially roots or stems—rather than with whole words. A given morpheme rarely occurs in only one word. Thus, elimination of a word does not necessarily remove the morphemes of which it is composed from the language. Even if a morpheme is lost from one language used in the reconstruction, the chances are exceedingly slight that that morpheme has been lost in related languages as well. Erosion is thus limited both by morpheme productivity and by the existence of related languages.

Yet another reason for morpheme persistence should be mentioned, even though it is as yet poorly understood. Limitations on semantic shifts are imposed by the cognitive requirements of lexical systems. Study of the nature of such systems is still in its infancy but it is an increasing concern in anthropology (for a historical approach cf. Berlin and Kay 1969), if not in linguistics, where it has been inhibited by the insistence of Chomsky on the primacy of syntax and the essential uninterestingness of paradigms.[2]

Any lexical system is represented by a series of interlocking paradigms, related through alignment of semantic features in relationships governed by analogy and opposition. Abstracted from the sequences in which they occur, the lexical items (morphemes, words) that bear these features are found to be systematically related to one another just as are phonemes, but as semantic rather than articulatory elements. A shift in the semantics of one item in such a system disrupts the pattern of the whole, often bringing in its train a reorganization of the entire system. The need to preserve system is thus both a barrier to, and a governing principle of, change. As successive stages of early language are isolated and relations between successive stages semantically analyzed, the nature of both resistance and reorganization will be better understood.

The conclusion of Saussure (1959) that the meaningful unit of language was *sign* rather than *symbol,* because of the arbitrary relationship between phonological representation and meaning, becomes questionable once a motivation is discovered for assignment of a particular meaning to a particular phonological unit. The present study shows that initially a spatial analogy between the manner of articulation and the mode of action or the relationship expressed by the meaning provided the symbolic motivation. While these initial analogies eventually became obscured through change, other profoundly motivating analogies became deeply embedded in the paradigmatic and syntagmatic interrelating of symbol to symbol. The assumption that the linguistic *sign* is unmotivated is based on too simplistic a view of the nature of language.

Analogy and opposition as the crucial attributes of the paradigm were recognized as important analytical tools in the Bloomfieldian structural tradition, under the rubric of *sameness or difference* in distinguishing *emic* from *etic* levels, fallen into disrepute since Chomsky 1957: 49–60) poured scorn on structural linguists for confusing *discovery procedure* with grammatical description. Chomsky has consistently failed to see that it is impossible to make a competent semantic feature analysis without an adequate paradigmatic exploration. Paradigmatic description must always underlie syntactic description. Even in as simple a structure as PL the organization of samenesses and differences between units has already reached an astonishing degree of complexity. As language expanded and changed, the semantic and syntactic base increased enormously in sophistication and intricacy, while the phonological base changed through time but failed to expand significantly. Separation of meaning from sound meant that phonological expansion became unnecessary.

Because Indo-European studies have carried reconstruction of PIE to a level not even approached for any other comparable family of languages, with almost total recovery of stem and inflective morphemes, problems that have been encountered in extending PIE reconstruction to a pre-PIE stage have been crucial to the reconstruction of PL. As recovery of PL phonemes progressed it became apparent that Hittite, unlike other Indo-European languages, possessed overt phonemic representation of almost the entire range of PL fricatives. These phonemes had coalesced with other Hittite phonemes, particularly stops. Morphemes containing these reflexes were so different in form from those in Indo-European, which had lost most fricative representation, that they had been assumed to be borrowed. Hittite was assessed as a heavily *mixed* language with strong substratum influences from other languages, especially Hattic. A considerable number of Hittite forms rejected by Indo-Europeanists as noncognate are easily seen to be cognate once the evolution of PL fricatives at pre-PIE levels has been recognized.

As the reconstruction progressed it became apparent that whole vocabularies rather than reconstructions must be used for languages or linguistic groupings other than Indo-European. This is because other reconstructions are partial and do not include instances of the morpheme productivity alluded to above. The possibility of borrowing can also be reduced if not eliminated where morpheme recurrence is found.

In general, languages that prove diagnostic (that is, conservative) with respect to certain phonemes or phoneme sequences are innovative for others. I have tested successive hypotheses on a wide variety of languages from most of the world's major stocks and some isolates, such as Basque in the Old World and Tarascan in the New. Early isoglosses emerged to link languages not previously classed together. Some reassessment of established groupings became necessary.[3] Indo-European was used as a base line in order to provide the greatest time depth available. Problems in PIE reconstruction became problems to be resolved in PL or in post-PL intermediary stages. Dictionaries became favorite bedside reading in the search for reconstructive clues.

The reconstructions presented in this paper represent only a small segment of the total PL thematic stock, both as reconstructed to date and as potentially reconstructible. Since *phoneme* and *morpheme* are isomorphic in PL, any discoverable sequence must be considered a complex stem or a thematic base to which inflective sequences were added. No effort is made in this preliminary study to assess the extent to which inflection existed at the PL level. A total reconstruction of PL sequences will necessarily be quantitatively greater than the inventory of reconstructed PIE roots and stems because not all of the former are accessible from the PIE data. There is also the possibility that some had been lost before emergence of the PIE horizon.

The bulk of the following material constitutes minimal, selected, nonrandom proof of genetic relationship between Indo-European, Hittite (H), Dravidian (D), Turkish (T) and Yurok (Y).[4] Indo-European and Hittite are shown to constitute a less firmly consolidated unit than has been previously hypothesized. Indo-European and Dravidian are families, or stocks, composed of a group of modern descendants. Turkish is generally placed within

Altaic, a grouping with doubtful reconstructive solidity. Yurok, a California Indian language, has been shown to be related within an Algonkian-Ritwan (Haas 1958) grouping uniting certain eastern and western languages of North America.

Inclusion of T presents a problem in that elimination of loan material from Arabic is extremely difficult if one is not an Arabic scholar. The problem is complicated further by the fact that the phonemic reflexes for Arabic from PL phememes are often identical in T and Arabic. I have struggled with the problem to the best of my ability with an Arabic dictionary and have often included more than one form from T so that those who know Arabic can make their own eliminations of Arabic forms if I have failed to do so.

Selection of these languages is, except for Indo-European and Hittite, more or less random. The language groups represented have previously been considered unrelated and adequate dictionaries are available. Selection of sets with form-meaning correspondence is nonrandom. Forms were selected from an Indo-European, form-alphabetized dictionary (Pokorny 1959) because they show initial contrast between all IE consonants, including hypothesized (now nonconsonantal) *laryngeals* before *e followed by a resonant or before a vocalic resonant. An effort was made to abstract stems from suffixal accretions and to include all such stems found in IE except those with meanings referring to light and sound. These were eliminated because, in most cases, metaphoric extension from more basic meanings can be detected.

The evidence presented is linguistic in the sense that it fulfills certain fundamental requirements of the comparative method. Whole words, or meaningful parts of words (morphemes—in this case stems) are compared for similarity of sound configuration coupled with similarity of meaning. From recurrent identity of sounds in each language in specifiable configurative environments (for example, before or between vowels) a hypothetical, prototypical sound in an earlier, parental language may be conjectured. Wherever such a sound occurred in a parental form its shape in daughter forms must be predictable. If either sound or meaning has changed beyond recognizability, reconstruction is clearly impossible. Both do, of course, undergo changes with the passage of time. There are no hard and fast rules for evidential admissibility. One criterion is quantitative, but the quantity of reconstructed forms that would be considered sufficient is unestablished. Another criterion is that meaning not be too divergent from the prototypical meaning. However, once solid sound reflexes have been established greater meaning divergency is considered admissible.

A more reliable criterion for the sureness of the reconstruction is, or should be, the degree to which the evidence is systematic, both on the parental and daughter levels. By this is meant the degree to which, for example, phonemes with a particular type of articulation on the earlier level develop into phonemes all (or most) of which are of a uniform articulatory type. Just as any phonemic transition is an articulatory feature shift rather than substitution of a completely distinct articulatory configuration, so semantic transitions eliminate or replace particular features rather than whole meaning configurations. New meanings, then, like new phonemes, are both similar to and different from those which they replace. Meaning glosses tend to obscure the similarities while stressing the differences. Meanings that seem quite different, when analyzed, often are revealed to have undergone little change.

As languages change, consonant clusters are the configurations that undergo the greatest modification. Because of this, in order to provide evidence of consonant contrast and demonstrate as clearly as possible the meaning of primordial consonants, the reconstructive canon was restricted to minimal stem configurations with any or no consonant in initial position and only a single resonant in final position, the two separated or not by a prothetic vowel, *e, which seems not to have had phememic status. This canon permits a variety of shapes: eR, CR, CeR, CRe, eCR.[5] Only rarely are reflexes of these shapes found as lexemes. Most frequently they are found as underlying stems, not necessarily still isolable as such, with an accretion of suffixal material that may vary considerably from language to language. To the extent to which they recur from language to language in the sample data the suffixal phonemes are reconstructed. The semantics of these suffixes is not discussed in the present paper.

The semantics of CR stems is primarily verbal and words belonging to other form classes are usually derived from such verbal bases. Verbal meanings are specific in one sense and abstract in another. Specificity derives from spatial movement or relationship, abstractness from the fact that such concepts as loosening or loss, for example, can underlie many specific actions or states, such as illness, death, destruction, weakening, resting, or untying. Gross, or dictionary, glossing tends to obscure underlying likeness of meaning while feature analysis tends to illuminate it.

The recovered stem meanings in PL are patently uncultural and far more basic than what is usually assessed as basic vocabulary by linguists. This applies to most PIE reconstructed vocabulary as well. Words for *rock,* for example, in daughter IE languages derive from stems with meanings of sharpness, projection and the like, or words for color from those of brightness or sheen. Cultural vocabulary must be assumed to be of far later date than the probably mid to late paleolithic horizon reconstructed here, when culture was incipient rather than formed.

Some graphic regularization of phonemic symbols in the demonstration languages has been made: H /ḥ/ becomes /h/; T /ʔ/ > /i̯/, /ç/ > /č/, /ş/ > /š/; Y /ɪ/ > /ə/; macrons are retained for vowel length in PIE, H and D but are the equivalent of a following raised dot in Y; PIE /i̯/ and /u̯/ are represented as /y/ and /w/ respectively.

Consonantal reflexes of PL phememes are presented in Table 2. This does not exhaust the consonantal inventories of these languages; H and T have an additional, voiced, stop system, while PIE had both a voiced and a voiced-aspirated series. These new phonemes arose from a morpho-phonemic reinterpretation of consonant clusters, especially fricative-stop sequences. These developments are not elucidated here since they fall outside of the stem canon under consideration. Sporadic reconstructed sequences of this type are occasionally provided without explanation.

The reflexes included in Table 2 are for all consonants in stem-initial position and resonants in stem-final position, H, T, D, and Y share hardening of front fricatives to stops. T, D, and Y share other features, such as stop nasalization and two to three separate reflexes apiece for *p, *t, *k^w, *m, and *n. Replacement of stops by nasals occurs sporadically rather than consistently in these languages, with T showing /b/ rather than /m/ as an alternate of *p. These phenomena probably do not reflect free variation but, rather, original fricative-stop clusters with initial fricatives that have not affected PIE or H stop or nasal articulation. Alternate reflexes have been presented here because it has not yet been determined which of the alternates represents the simple, consonantal initial.

Variation in Y between /l/ and /r/ as a reflex of *r is probably attributable to the secondary development of *smallness* as a meaning feature of /r/ and *largeness* as a feature of /l/, resulting in word pairs differentiated by an /l/, /r/ alternation (Robins 1958:14). Y /ʔ/ is a reflex of *š, but initial /ʔ/ also developed as a prothetic consonantal onset wherever that of PL was vocalic.

TABLE 2
Consonantal Reflexes

PL	PIE	H	T	D	Y
p	p	p	b, h	m, p	m, p
t	t	t, z	n, t	n, t	n, t
c	ṣ	š	c	c	c
č	k	k	č	t	k
k	k	k	k	k	k
k^w	k^w	k, ku	g, k, ku	k, ku	k^w, k, ku
f	ø	p	ø	p	p
θ	ø	t	t	t	t
s	s	š	s	ø	s
š	ø	ø	s	ø	ʔ
x	ø	h	ø	ø	h
x^w	ø	h	ø	k	k^w, k
m	m	m	m,b	m, ø	m, ø
n	n	n	y, ø, n	n, ø	n, l
l	l	l	l	l, n	ł, l
y	y	y	y, ø	y	y
r	r	r	r	r	l, r
w	w	w	v	v	w

*šV metathesizes in Y, becoming, in initial position, /ʔVʔ/, so that a contrast between the original *š initial and V initial is maintained. Y shows a greater degree of metathesis in general than the other languages. It is probably conditioned rather than random, but the parameters of the phenomenon have not yet been determined.

In Hittite, medial voiced stops were written as single, and voiceless stops as geminate, consonants. H /z/ reflects a PL *ty or *ts sequence (cf. Sturtevant 1964).

All of the languages show at least some weakening and loss of fricatives, and all but PIE show some hardening of fricatives. D and Y show the greatest uniformity in this regard, indicating a shared history beyond the split-off point of T.

As far as has been determined T, D, and Y all reflect the original stop or nasal in medial position, although this may be lost in reflexes of some PL clusters. In D, articulation of medial dental stops and nasals is also modified by following consonants that have been lost. Probably these were largely or wholly fricatives.

Since analysis of consonant clusters is not germane to this paper, no cluster reflexes are included in the Tables. They do occur sporadically, however, in the suffixal increment in certain lexical sets presented as reconstructive data. Briefly, *xʷS is reflected as an aspirated, voiced stop in PIE and as a simple, voiced stop in H and T. *fS is reflected as a voiced stop in H and PIE and *xS as a voiceless, aspirated stop in some IE languages. *sS is reflected as *(s)S in PIE but as S in H. Prestop *θ and *š seem to have been lost altogether in PIE and H. D and Y retain traces of medial clustering in some environments.

Fricatives, resonants, and *kʷ all had both consonantal and vocalic allophony in PL. For fricatives and *kʷ the vocalic allophones had a consonantal onset in certain environments, especially after word juncture. Vocalic reflexes of *kʷ are included in Table 2. Those of fricatives are shown in Table 3, and those of resonants in Table 4. These contrast in that fricatives show consonantal onset and resonants show consonantal offglide.

TABLE 3
Syllabic Allophones of Fricatives

PL	PIE	H	T	D	Y	
f = [fö]	ə	pā	ö	po	pə	24, 25, 28, 41, 44
θ = [θa]	e[C	(ta)	ta	ta	ta	6, 29, 42, 95
s = [sə]	ə	šā	si	a	sə	1, 4, 61, 71
š = [si]	ə	ā	si	a	ʔə	4, 59, 78, 81, 84
x = [xü]	ə	ha	ü, ö	o, a	hu, ha	1, 3, 8, 58, 65, 66, 68, 99, 100
xʷ = [xu]	o[C	hu	o, u	ko, ku	kə, ka	2, 9, 15, 49, 70, 79, 91, 94

88

TABLE 4
Syllabic Allophones of Resonants

PL	PIE	H	T	D	Y
m = [m̥]	m̥	a(m)	a(m)	a(m)	o(m)
n = [n̥]	n̥	a(n)	a(n)	a(n)	o(n)
l = [l̥]	l̥	a(l)	a(l)	a(l)	o(ł)
y = [i]	i	i(y)	i(y)	i(y)	i
r = [r̥]	r̥	a(r)	a(r)	a(r)	o(l), o(r)
w = [u]	u	u(w)	u(v)	u(v)	u(w)

Turkish shows a complete range of contrastive vowels representing the original allophony. I hypothesize the same, or similar, range for PL, shown in square brackets in Table 3. Again, the variation in reflexion of the velar fricative in T, D, and Y is a shared, and as yet unexplained, isogloss.

Just as juxtaposition of fricative and stop gave rise in PIE to consonant coloration, so juxtaposition of certain fricatives and *e gave rise to PIE vowel coloration. Once the coloration was established as contrastive, the fricatives became weakened and eliminated. Such a development is exclusive to PIE, although in the other languages vocalic allophones of fricatives may be backed, fronted, raised, or lowered in response to their consonantal point of articulation, as shown in Table 3. Fricative + vowel vocalism is shown in Table 5.

The evidence from Hittite is somewhat limited, hence inconclusive. It seems to indicate that Hittite showed none of the backing and rounding influence from a preceding fricative found in PIE vowels. The evidence seems to me better for postulating H /a/ as a vocalic reflex of liquids and nasals than /u/ as Sturtevant (1964) hypothesized. In the limited data presented here syllabic fricatives, except velars, seem to result in / ā/ vocalism (cf. Table 3).

In contrast to PIE vocalic rounding and backing in sequences of rounded and/or back fricatives + *e, notice T, D, and Y vocalic raising in

TABLE 5
Fricative + *e Reflexes

PL	PIE	H	T	D	Y	
fe	o	pe	e, i	e, i	e, (iʔ)	24, 25, 28, 41, 44
θe	e	te	te, ti	te, ti	te, ti	6, 23, 29, 42, 95
se	se	še	se, si	e, i	se, si	46, 61, 71, 101
še	a	(aʔ)	se, si	e, i	ʔe, ʔi	4, 59, 78, 81
xe	a	he	e; a[w, y	e	he	1, 3, 8, 58, 65, 66, 68, 99, 100
xʷe	o	(?)	e	ke, ka	ke, kʷa	2, 9, 15, 49, 70, 79, 91, 94

sequences of labial, dental, alveolar and palatal fricatives + *e. No determining factors for this phenomenon have been discovered to date. It mav tentatively be assigned to free variation.

Fricative coloration of PIE vowels has unfortunately come to be known as the *laryngeal hypothesis,* although the original theory, proposed by Saussure (1879), named the lost phonemes only *coefficients sonantiques* without specifying their articulatory characteristics except that they were realized in different environments with either consonantal or vocalic allophony, just as PIE resonants. Consonantal allophones had been lost before and after *e, leaving backing and/or rounding as residue in some cases. *Coefficients sonantiques* became transmuted to laryngeals after Kuryłowicz (1927) pointed out that Hittite displayed /ḫ/, a velar or laryngeal spirant, in some sequences in which the missing phonemes were to be expected. Since that time, vanished laryngeals have varied in number with the individual theorist, with three to six the most common range. The Hittite evidence was confusing, as /ḫ/ occurred sporadically when other IE languages showed reflexes for PIE *a, *o, and *e. Puhvel is also puzzled because H frequently shows initial prevocalic or preconsonantal /hu-/ in the absence of other clusters of initial laryngeal plus continuant (Puhvel 1965:88). I am, in turn, puzzled that a systematic solution did not easily suggest itself, since the PIE system also contained *k^w, which occurs similarly as /ku/, both prevocalically and preconsonantally.

When it was shown that Hittite was related to IE, it was assumed that the former was a very mixed language, with a large number of borrowed forms, because many words seemed not to reconstruct to a PIE model. The discovery that pre-PIE laryngeals were originally fricatives and that they were of the same number and points of articulation as both the PL stop and resonant systems suggested to me that a new look at the Hittite lexical residue might be in order. Some of the results of ʳhis reevaluation are given in the present paper. This evidence shows Hittite to be not just another IE dialect, which has been increasingly insisted upon by some Indo-Europeanists (cf. Puhvel, 1966), but divergent enough so that its position in the sequence of changes and branches that took place between the time of PL and the separation of PIE must be carefully assessed.

*e is probably best explained not as a phememe like the consonants but as a prosodically prothetic addition responding to stress considerations.[6] Vocalization of resonants and fricatives also seem to be conditioned by patterns of stress and syllabification that are not yet determined.

There are sporadic problems in the following sets (see below) that will ultimately need to be resolved. One such problem is the origin of long vowels in D and Y. No attempt will be made to deal with this here.

The material on which Tables 1–5 are based is presented in the sets of cognate forms that follow. The capitalized glosses represent some sort of semantic common denominator rather than original meanings of the reconstructed stems (preceded by asterisks). The original meanings are included in parentheses after the gloss and themselves represent a semantic condensation of the features shown in Table 1. Each phememe is composed of a bundle of five features, four of them hierarchically arranged to show degree of sharing and differentiation.

RECONSTRUCTIONS AND COGNATE SETS

1. ANCESTOR, ORIGIN (oppositive-interpositive)   *x(e)n-
   (also *x(e)nt-, *x(e)ns-, *x(e)nf-, *x(e)nx^w-):

   PIE   *an- *grandmother, grandfather,* *anti *opposite, before,* *ant-s *front,* *andh- *sprout, bloom*

   H   hannaš *grandmother,* haš- *procreate, bear,* hanza *front* hamešhant- *spring (of year)*

   T   ana *mother,* anne *mother, maternal grandmother,* an'ane *tradition,* anasïl *by origin,* ensal *generations,* ön *front, the future,* önce *first,* önüsïra *before him,* öte *the farther side*

   D Ta attan *father, elder,* aṇṇā *elder brother, father,* aṇai *copulate with, approach,* accan *father,* okkal *relations, kinfolk,* ōmpu *protect, cherish, bring up*

   Y   hunowom(c-) *bring up, rear*

2. APERTURE, ANGLE (apertive-interpositive)   *x^w(e)n-
   (also *x^w(e)ns(e)r-, *x^w(e)nft-):

   PIE   *oner- *dream,* *ond-, *ṇd- *stone, cliff*

   H   hantī *severed, sundered*

   T   engebe *unevenness of ground, broken ground,* ense *nape of neck, back,* unf *roughness,* unk *neck*

   D Ta kēni *small tank, well, ditch, trench,* kan *eye, aperture, orifice, joint in bamboo* or *cane,* kāṇ *see,* kaṇṭi *gap in hedge* or *fence, mountain pass,* kanavāy *mountain pass, ravine,* kuṉṟam *hill, mountain,* kūn *bend, curve*

   Y   knuʔlogeł *to be deep,* kʷereʔwey- *to have a pointed face, sharp tongue*

3. APPORTION (oppositive-horizontal)   *x(e)y-
   (also *x(e)ynk^{w}s-):

   PIE   *ai- *give, share*

   H   hink- *apportion, hand over*[7]

   T   ayal *families, dependents,* ayniyar *goods in hand, inventory* ayïr•mak *separate, choose, distinguish,* inkïsam *being divided*

   D Ta aiyam *alms,* oy *send forth, give,* iku *give*

   Y   heyomoks- *to be lucky*

4. AROUND, LIKE, BOTH (vertical-bilateral-apertive-projective)
                                             *š(e)mx^{w}p-

   PIE   *ambhi, *ṃbhi *around, on both sides*

   T   sevabïk *precedents, antecedents,* sevab *God's reward,*

   D Ta a-ppāl, a-ppuṟam *that side, beyond,* i-ppuṟam *this side*

   Y   ʔomoki *in return*

5. BEGIN, NEW, BIRTH (interpositive-circumscriptive)   *n(e)w-
    (also *n(e)ws, *n(e)wsxʷec):
    PIE   *newos, *newyos *new*
    T   yeni *new, recent*, yavru *young of bird or animal*, yuva *nest, home*
    D Ta niva *to rise, grow*, nivar ~ ivar *to rise on high*
    Y   nes(kʷec-) *to come, arrive*

6. BEND, FORK (juxtapositive-extenuative)   *θ(e)l-
    (also *θ(e)l(e)θ(e)y-, *θ(e)l(e)y-):
    PIE   *el-, *elēi- *bend*
    T   tel *wire, fibre, a single thread* or *hair*, talaz *wave, billow*, talaš *wood shavings, sawdust filings*, talik *suspension, making a thing depend upon something else*, alkïm *rainbow*
    D Ta tāḷ *leg, foot, stem, stalk*, talai *head, top, end, tip, hair*, taḷir *to shoot forth, sprout;* Malt téle *to lift, as the corner of a curtain* or *the hem of a dress*
    Y   teltelun- *to be branchy, twiggy*

7. BEND, SWAY, FLEXIBLE (extenuative-interpositive)   *l(e)n-
    (only found extended, as *l(e)nk-, *l(e)ntxʷ(e)-):
    PIE   *leng- *bend, sway*, *lenk- *bend*, *lento *flexible, yielding*
    T   lenk *lame*
    D Ta āṭu *move, swing, shake*, natuṅku *to shake, tremble, falter, waver;* Pa ēnd- *to dance*
    Y   lenewkʷ- *to drift*, łkekʷol- *to limp, be lame*

8. BEYOND, FAR FROM (oppositive-extenuative)   *x(e)l-
    PIE   *al *beyond that*
    T   el *one other than oneself, those outside one's family*, eloğlu *stranger, outsider*, öl- *death*
    D Ta al *that time*, alo•n, alo•l, alo•r, ald *that previous person*, ali•k *to the previous place*, ellai *limit, extremity, goal*
    Y   hełkew *in the mountains*, hełku(s) *ashore, on land*

9. BOUNDARY AROUND APERTURE (apertive-circumscriptive) *xʷ(e)w-
    PIE   *ou-tlā *bandage, wrap*, *outo-s *clothed*, (*eu- *to put on*, cf. 30), *əus-, *ōus- *mouth, estuary, shore, car*, *owi-s *sheep*
    H   aiš *mouth;* Luw hawi-, Hierogl. Luw hawas *sheep*[8]
    T   avaz *loud voice*, avret *genitals*, avlu *courtyard*, ova *grassy plain, meadow*, avrat *wife, woman*, avuc *palm of hand*
    D Ta cevi *ear* (c- < PD *k-) kevi *deep valley, cave;* Kol kopli *mouth*
    Y   kəwəyəh *chin, jaw*, kʷeykʷeyur, kʷəhkʷəyəc- *to whistle*, kewoy *burden basket*, kowištewoł *cemetery*, rekɔʷoy *river mouth*, rik'ew *shore*

10. CARE FOR (intromissive-circumscriptive)   *t(e)w-
    (also *t(e)wtš-, *t(e)wtxʷ-):
    PIE   *teu- *to be friendly, vigilant, watchful*, *teutā *folk, land*, *teutono-s *ruler*
    H   tuzzi *lord, encampment*
    T   tevakki *a taking care of oneself, being on guard against*, tevdi *committing to the safekeeping of another*, tavzif *entrust*, nevzad *newly born (child)*

D Ka tevalu, tevulu *an itching desire, an inordinate attachment;* Ta nampu *long for, trust, confide in, rely on,* etc., tuṇai *association, help,* etc.

Y newonoc *to suckle,* nowk^w- *to care for*

11. CENTRIFUGAL (emissive-circumscriptive) *c(e)w-
(also *c(e)wš-, *c(e)wx^w-, *c(e)wkrx^w-):
PIE *seu-, *sū, *sewə-, *sū-bend, *revolve, impel;* *su-k-ro *twisted*
T cevelân *revolution, circuit, circulation,* coš- *boil,* exuberance, civelek *brisk, lively, playful*
D Ta curi *whirl, curl, be spiral,* cuṟal *whirl, spin, rotate,* cuṟṟu *to revolve,* etc., cukir *to card as cotton, separate as hair, fibres;* Ka cuñcu *the hair curling round the forehead*
Y curḃay *to comb*

12. CLEANSE, WIPE, PURIFY (projective-circumscriptive) *p(e)w-
(also *p(e)w-s, *pw(θ)-rx^w(e), *p(e)w-tx^w(e)):
PIE *pew-, *pewə-, *pū- *to clean, purify, filter,* *pū-ro-, pŭ-to-*clean*
T havlu *towel,* huzu *humility,* huzur *repose,* hutbe *Friday prayer*
D Ta mavvam *beauty,* muṟuku *to bathe the entire body by dipping or pouring, be immersed;* Te musuru *a constant or continued rain,* muka *to draw water*
Y meworoh *to be clear weather,* meworoy *to flow away,* mewoletew- *to wipe one's hands,* mewoleʔwey *to wipe one's eyes;* mulonem- *to wipe,* muntəʔəy(-) *to be white*

13. COMPLETE (emissive-interpositive) *c(e)n-
(also *c(e)nw-, *c(e)nx^w-):
PIE *sen-, *sene-, *sen(e)u-, *senə- *prepare, perfect, complete,* achieve, *sen(o)- *high,* *seno-māter *grandmother*
H šanh- *seek, strive for, further*
T cenab *majesty, excellence,* cennet *paradise,* canatiš *passionate desire,* can *darling, beloved friend*
D Ka cannu, cennu *straightness, beauty, grace, niceness, properness, elegance;* Tu sāṅkuni *to foster, nourish, nurse, bring up, take care of, protect, shelter*
Y cewonem *to straighten,* -cekos *mother*

14. CONCAVE, CONVEX (clinal-circumscriptive) *č(e)w-
(also *č(e)wš-, *č(e)wx^w-):
PIE *k̂ew-, *k̂ewə-, *k̂ū-, *k̂wā- *swell, swelling, arch cavity, concave, convex*
T čevirmek *turn, turn round,* čevre *circumference, circuit, surroundings, contour,* čukur *hole, hollow, ditch, cavity, tomb, dimple, hollowed out, sunk, concave,* češm *eye*
D Ka tūtu, tūnto *hole,* tumbu *to become full, complete, plump, strong;* Ma tēvuka, tēkuka *to draw water, empty a well, bale out for irrigation*
Y cf. HOLLOW

15. COPULATION, INSERTION, EMISSION (apertive-horizontal) *x^w(e)y-
(also *x^weyx^wp-, *x^weyft-):
PIE *oi- *pole, rod,* *oibh- *to fuck,* *oid- *to swell,* *oidos *tumor*
H huinu- *cause movement or flow, move forward,* huittiya *to pull*

T   oyuk *hollowed out, gouged out,* oylum *excavation, pit, hollowed out,* oymak *excavate, scoop out, carve,* idiš, idič *gelding, castrated*

D Ta   koy *pluck, cut, reap, shear,* cī *pus, snot;* kuyam *sickle, reaping-hook, curved knife, razor;* Te koyya *stick, rod, staff;* Ka kaidu, keydu *weapon;* Ma kaya̱ruka *to increase, rise, ascend,* kumpi *penis*

Y   kə?yəwet *to release,* kəycəł *ring-tailed civet cat,* ke?yolew *to spit,* ke?yonem *to release*

16. CRUSH, BREAK (cf. GRIND) (extenuative-bilateral)   *lem-

PIE   *lem- *break, broken, weak*

T   melhame *battle, carnage*

D Ka   namalu, nevaru *to chew, masticate, munch, chew the cud,* name *wear away,* etc.; Ta amuṇku *crush, press*

Y   łmey *bad, mean, nasty*

17. CUT, CHEW, SEPARATE (intromissive-bilateral)   *t(e)m-
    (also *t(e)mx-, *t(e)mnf-):

PIE   *tem- *to cut;* (Gr. temnō)

T   temyiz *a separating* or *distinguishing,* namahrem *not related* or *intimate, not having access to the harem*

D Ta   nēmpu *to winnow,* tēmpu *to fade, wither, droop, be tired, faint, grow thin, be emaciated, be in trouble, suffer, perish;* Te nēmu *to winnow, sift*

Y   tmoh *half,* tmohkeloyt *to break,* nohsec- *to take pieces off, chip*

18. DECAY, DISTEND, WEAKEN (extenuative)   *(e)l-

PIE   *el- *decay, hungry, bad*

T   elem *pain, sorrow, illness,* alčak *low, vile*

D Ta   eḷimai, eḷumai *despicableness, lowness of rank, circumstances* or *character, poverty, weakness, depression of spirit,* ala *to suffer, be in want,* alar *to blossom, open up,* alkul *side, waist, pudendum muliebre*

Y   łk *dirt, slime, mud*

19. DEFICIENT IN GROWTH, FAIL (extenuative-sublative)   *l(e)r-
    (also *l(e)rfk-, *l(e)rft-):

PIE   *lerg- *smooth, even, slippery,* *lerd- *curve*

T   lerz *a tremble, shiver*

D Ta   naruṅku *be deficient in growth, decay, grow lean as a child, fail as a business* or *harvest;* Kur nerr *snake;* Ta nēr *to grow thin, lean, be emaciated, soft, yield to the touch, slenderness;* Ma nērkka *to become thin, fine,* nērppu *fineness, thinness, liquification;* Ta a̱ri *to perish, decay, fail,* etc.

Y   lełko•? *to fall*

20. DELAY, STAY, SUPPORT (intromissive-extenuative)   *t(e)l-
    (also *t(e)l-s and possibly *ty(e)l- or *ts(e)l-)

PIE   *tel-, *telə- *carry, support, endure, be steadfast*

H   zaluganu *delay, pull out in length*

T   toloz *arched vault,* tel *wire, fiber*

D Ta   nil *to stand, stop, halt, be steadfast*

Y   no•ł adv., sent. con. *then, far, long*

21. DIE (bilateral-sublative)[9] *m(e)r-
    (also *m(e)rs-, *m(e)rsx^w-):

    PIE   *mer-, *merə- *die;* *mr̥-tó-m, *mr̥-ti, mr̥-tú *death,* *mr̥-wó, *mr̥-tu-o- *dead*

    H   maršah- *decay, mark- decompose*

    T   berbad *destroyed, scattered, ruined, spoilt*

    D Ta ari *to perish, suffer, be used up, destroy,* etc.; Ma mār̠kuka *to languish, grow faint, sleep, die*

    Y   merkewec- *to die, faint,* merk^{w}ewoł *to perish,* moyk- *to die (gods, plants, animals)*

22. DIG, HOLE (delative-interpositive) *k(e)n-

    PIE   ken *scratch, rub*

    T   kende *dug, excavated, engraved*

    D Te kannamu *hole, bore, orifice, hole made by a burglar in a wall*

    Y   knuʔloget *to be deep*

23. DIG UP, PLOUGH (juxtapositive-sublative) *θ(e)r-
    (also *θ(e)r(e)fp-):

    PIE   *ereb- *bore, use sharp tool, hollow out*

    H   terippi- *ploughed*

    T   terip- *plough,* tarak *comb, rake, harrow,* ara *interval, space,* ark *irrigation canal,* tirpidin, tirpit *small mattock*

    D Pa terip *to churn;* Te t(r)avvu *to dig, excavate, scoop, unearth, stir up;* Ka tari *to be chafed, abraded* or *grazed;* Ta tira *to open*

    Y   tik^{w}ohs- *to break* (tr.), tik^{w}oł *there is a canyon, low gap*

24. DISPARAGE, DISPLACE (abessive-interpositive) *f(e)n-
    (also *f(e)nx^w(e)-):

    PIE   *ono-, *on-, *(o)no-d- *abuse, belittle*

    H   pangariya *get the upper hand*

    T   önle•mek *resist, stop, prevent*

    D Ta pani *be humble, become inferior,* piṇam *corpse, carcass, disembodied* soul, piṇi *disease, sickness, suffering,* poṭi *dust, ash, particle, become broken* or *pulverized,* poṉṟu *perish, die, fail, be ruined,* pōkkaṇam *shame*

    Y   penk^{w}eł *to be eaten hollow by bugs,* pontet *ashes,* pəncəc *dust*

25. DRESS, SURROUND (abessive-circumscriptive) *f(e)w-
    (especially *f(e)wt-)

    PIE   *outo-s *dressed*

    H   putalliyant- *lightly clothed*

    T   utan•acak *shame, modesty,* cf. 30.

    D Ta puṭṭil *quiver, sheath, basket,* purai *house, dwelling*

    Y   pu•wiš *sack*

26. DWINDLE, WEAKEN (extenuative-horizontal) *l(e)y-

    PIE   *lei- *decay, take off, dwindle, thin,* *lei-bhō, *lei-no- *weak,* *lei-ro- *slender*

T   liynet *looseness of the bowels* (arabic ?)

D Ta iḷai *grow weary, be fatigued*, etc., nil *to stand, stop, stay, cease, delay*, etc.

27. EMIT, FLOW, SUPPURATE (sublative-horizontal)   *(e)r(e)y-
(also *(e)ryx^w-):

PIE *erei- *flow*, *rī-, *rī-ti *flowing*

H   arha *forth, from*, arhaya(n) *separated* (cf. 80)

T   ihrac *extraction, expulsion, emission*, irin *pus*, irlenmek *to suppurate*

D Ta eruku *to have loose motions (said of cattle)*

Y   royk- *to flow*, ri·k(omoy-) *to be full*

28. EMPTINESS, DECAY (abessive-extenuative)   *f(e)l

PIE *ol- *decay*

T   alil *ill, invalid, blind*, ölüm *death*, ölmek *die, fade, wither*, illet *disease, defect*

D Tu paḷeṅkuni *to upset, overturn;* Ta paḷḷam *lowness, lowland, valley, ditch*, etc., pollu *empty husk of grain*, piḷ- *to burst open*, etc., piḷa *to be split, gape*, etc.

Y   poɬk^wen- *be moldy*, pəɬk^w- *to be grey*

29. EMPTINESS, LACK, CESSATION (juxtapositive-circumscriptive) *θ(e)w-
(also *θ(e)wx^w-, *θ(e)wθw):

PIE *eu-, *euə- *lack, empty*

H   tuhhuš-, tuhš- *cut off*, tuhšant- *cut off, shortened*

T   tevakkuf *a stopping, tarrying*, tevlid *a giving birth*, tevessu *a being spacious* or *extensive*, tevehhüm *an imagining, fancying, groundless apprehension* or *foreboding*, tavik *to hinder, delay, prevent*, tutuk *paralyzed, stopped up, impeded*, taviz *substitution*, tavaši *eunuch*

D Ta tava *to cease*, taval *diminishing, decreasing, failure, death, poverty*, tukku *meanness, worthlessness*, tuṇi *be sundered*, tumi *to be severed, obstruct;* Ka tūtū *hole*

Y   ta· *not at all, nothing at all*, tohpew *hole*

30. ENCLOSURE, ENCLOSE (circumscriptive)   *(e)w-
(also *(e)wf-):

PIE *eu- *put on clothing*

H   unuwā(i)- *decorate, adorn* upnu *handful*

T   ev *house, dwelling, home, household, family*, evani *vessels, pots, dishes*, evlek *furrow*, evlenme *marriage*, ufûl *setting (of the sun, etc.)*, evliya *guardians*

D Ta vēy *to cover (as a building), roof, thatch, put on (as a garland);* vētu *cover for mouth of a vessel*, vēḷ *to marry*, ū *flesh, meat*, ūṭu *middle, waist, that which comes between*

Y   (ʔ)wegah(pemew) *to get married*, (ʔ)wesk^weɬ *member of a family*, weʔyes- *to buy a bride*, ʔupur *raised ridge round basket*

31. EXTRACT (emissive-extenuative)   *c(e)l-
(also *c(e)l-w):

PIE *sel-, *sel-wo- *take, grasp*

T   celb *a procuring*, celbeder *attract, bring*, câlib *attracting, causing*

D Ka seḷe *to draw, pull, pull off, rob;* Ta ceḷḷu *flea, tick,* caḷḷu *to slacken, abate*
Y   cuł *well, goodby!,* cowon- *to fish with a seine net*

32. EXTRACT, EMIT (BLOOD, ESSENCE)

(sublative-circumscriptive)   *r(e)w-
(also *r(e)wx^w-):

PIE   *reu-, *rewə-, *rū- *burst, dig, grub up, pull up, gather up;* *reudh- *red*
H   ešharwant- *red*
T   ravza *garden,* revac *a being in demand,* revan *going, flowing, current,*
      revnak *brightness, splendor,* ruh *the breath of life, soul, spirit, energy,*
      activity
D Ta uravam *strength, force,* uru *to assume a form, issue forth, appear, come into*
      existence, body, shape, form, beauty, idol, urupu *form, shape, color,* uṟu *to*
      plough, dig up, root up, uṟavu *ploughing, agriculture,* ūr *decay, rot,* ūṟu *to*
      spring, flow, ooze
Y   reweyet- *to tow,* lewoleyt- *to pull a boat through shallow water by hand,*
      lewet- *to fish with a net,* roˑwos- *to smoke a pipe,* roˑk^w *wind,* roˑk^ws- *to*
      blow (of wind), be flatulent

33. FEED (projective-interpositive)   *p(e)n-

PIE   *pen- *feed, food, nourishment*
T   hanim *lady,* han *inn,* hane *house, dwelling*
D Ta peṇ *female,* maṅkai *woman,* moñci, moṇṇi *breasts*
Y   nep- *to eat;* menok^wolum- *to gulp down;* penk^w *acorn flour*

34. FILL, FLOOD (projective-extenuative)   *p(e)l-

(also *pelx-):

PIE   *pel-, *pelə- *fill, flow;* pḷ-tó *full*
H   palhašti-, palheššar *breadth*
T   halic *strait, estuary,* hayli *much, many,* balik *fish*
D Ta mali *to abound, be full,* malir *to flood,* poti *fullness, stoutness,* poli *to*
      flourish, abound, poḷḷu *to blister, swell*
Y   plohp- *to be in spate, flood,* moʔohkeloy *to be large and round,* moʔohkeroy
      to be small and round, plo- *to be wide,* pləwən(əy-) *to be high tide*

35. FLOW I (emissive-horizontal)   *c(e)y-

(*cs(e)y- or *cš(e)y-?):

PIE   *sei- *drip, flow, damp,* *seik^w- *pour out, strain, flow, drip,* *seip-, seib-
      pour, strain, flow, drip, *seik-, *seig- *urinate*
H   šiptamiya- *a flowing*
T   seyelân *a flowing, flood,* seylâb *flood, torrent,* seyyal *fluid, liquid,* sidik
      urine, cɨvɨk *wet, sticky, viscid*
D Ta ciṇukku *ooze, drip, drizzle;* citar *raindrop;* cilucilu *to rain gently,* cēkku
      breast milk, cīntal *mucous of nose*
Y   (ʔ)yoc *boat,* siyow *to break through waves (of a boat)*

36. FLOW II (originally 'spring'?) (emissive-sublative)   *c(e)r-

(also *c(e)rx^w-, *c(e)rš-):

PIE   *ser- *flow,* *sero-m *flowing,* *ser-mo- *river,* *sṛ-, *srə- *stream, flow*

T   cereyan *a flowing,* cerahat *suppurating,* cari *running, flowing,*

D Ta   cari *to slip away, slide down, slant, slope, pour down,* cāṟu, cēṟu *juice, sap,* cōr *to trickle down as tears, blood* or *milk,* cori *to flow down*

Y   coʔworec *driftwood*

37. **FORM, CREATE** (circumversive-sublative)   *kʷ(e)r-

PIE   *kʷer- *to make, form*

T   -ger *maker, worker, doer,* kur- *establish, plan*

D Ta   karu *foetus, embryo, egg, germ, young of animal,* karuppai *womb,* karu *mould, matrix,* karuvi *instrument, tool,* kuṟi *to design*

38. **FURTHER, FORWARD, EARLIER** (projective-sublative)   *p(e)r-
(also *p(e)rxʷ-, *p(e)r(e)x-)

PIE   *per- *danger, experience, effort, risk,* *pero-s *further,* *perā- *earlier*

H   parā *forward, further,* parh- *hunt, pursue, let gallop*

T   hercai *ubiquitous, roving,* hareket *movement, act, behavior, departure,* haric *the outside, exterior, abroad,* berayi *for,* bereket *abundance, increase*

D Ta   maṟu *another, other, next, beyond,* maṟa *youth, infancy,* maram *valour, bravery, war*

Y   mela *past time,* merogeyah *it is long since,* poy *ahead, in front*

39. **GO I, MOVE HORIZONTALLY** (horizontal)   *(e)y-

PIE   *ei-, *i- *go*

H   iya- *go*

T   eylemek *do,* iyadet *visit,* yiv *groove, stripe,* yel *wind,* yelmek *run,* yol *road, street, canal, channel,* yol- *send, travel*

D Ta   ey *to discharge arrows,* eytu *to approach, reach, obtain,* iyaṅku *to move, stir, go, proceed, walk about*

Y   yeguʔuh *ferryman,* yeʔw(omey-) *the sun sets*

40. **GO II, MOVE ABOUT** (expansive-horizontal)   *s(e)y-
(also *s(e)yc-, *s(e)yxʷ-):

PIE   *s(e)y- *let fall, throw, send, reach*

H   šāi- *go*

T   seyir *movement, progress, travel,* seyahat *journey, travelling, expedition,* isdar *issue, put forth*

D Ta   ēku *to go, pass, walk,* acaṅku *to stir, move, shake,* ayakku *to shake*

Y   sekoyor- *to run quickly,* so•yc- *to go quickly,* skewic- *to go slowly, go easily,* sekin- *to make an effort*

41. **GO FORWARD, DEPART** (abessive-horizontal)   *f(e)y-
(also *f(e)yw-):

PIE   *oi-, *oi-wo *go*

H   pāi *go*

T   ayak *foot,* ayri *apart, separated,* eylemed *do,* eyyam *days,* evgin, ivgin *hurried,* iyadet *visit*

D Ta   peyar *to move, depart,* pō (pōv-) *go, proceed, die, etc.*

Y   pyekceni *for a little while, carefully,* pyekcoh *to approve,* pyurker *to play,* peycew *deceased grandfather,* pyuc *well, all right*

42. GRASP (juxtapositive-bilateral)   *θ(e)m-
    (also *θ(e)m(e)s-, *θ(e)m(e)š-):
    PIE   *em-, ₑm-, *ₑmā, *ₑmē- *take*
    H   tamaš- *press*
    T   temellük *to take possession of,* temas *contact,* temeššük *a taking firm hold,* temin *a making safe, sure,* temhir *a sealing,* timsal *image, picture, model*
    D Te tami *love, desire,* tevvu *to get, take, obtain*
    Y   tmeg *to hunt for game*

43. GRIND, CHEW (bilateral-extenuative)   *m(e)l-
    (also *m(e)lxʷ-):
    PIE   *mel-, *melə-, *mlē-, mela-k = mlāk- *grind, mill, hit,* *ml̥-to-, ml̥-to- *crushing, pulverization*
    H   malla *grind*
    T   melhem *ointment, salve,* mahlut *mixed, adulterated, mixture*
    D Ta mel, melku *to chew, masticate,* malaṅku *to be agitated, turbid,* māl *to be confused, perturbed,* māl̤ *to be exhausted, expended* or *finished,* al̤, al̤am *sharpness, keenness of edge*
    Y   mel̤ownem- *to touch,* ʔelkel̤ *clay*

44. HEIGHT, ABOVE (abessive-sublative)   *f(e)r-
    also *f(e)r-w, *frtxʷ(e), *fr-s):
    PIE   *or- *rise, raise,* partic. *r̥-to, *or-meno, *orew- *fat*
    H   pargatar *height,* peruna *cliff, crag, rock*
    T   ergin *mature, ripe, adult,* uruc *ascent, ascension* arš *Throne of God, roof, pavilion,* ört•mek *cover,* örtü *cover, roof,* iri *huge,* irtika *a rising, ascending*
    D Ta paraṇ *watchtower, upper storey, loft,* paru *to become large, bulky, swell,* porai, porṛai *mountain, hill,* pŏr *cover, heap,* poṛutu, pōṛtu, pōtu *sun, time, opportunity*
    Y   perey *old woman,* peʔr, pl. pegerey *to be big (houses),* pelil̤ *to be big (human),* ploh- *to be big (body parts, round things),* pləwən (əy-) *to be high tide,* pləʔəy(-) *to be big (animals, birds)*

45. HELP, BE NECESSARY, DESIRE (circumscriptive-interpositive) *w(e)n-
    (also *wenθ-):
    PIE   *wen-, *wenə- *support* (> *wish, demand,* etc.)
    H   wen-, went- *future*
    T   unarmak *mend, set to rights,* unsur *element, root, component part*
    D Ta vēṇṭu *to want, desire, request, be indispensable, necessary*
    Y   wen- *to come*

46. HIGH, RAISED, AFFIXED (expansive-sublative)   *s(e)r-
    (also *s(e)rxʷ(t)-, *s(e)rft-, *serx-):
    PIE   *s(e)r- *align, affix*
    H   šarā *on,* šarazzi *above,* šarlāi *raise,* šēr *above, on*
    T   ser *head, chief, top, end,* sïra•daǧ *mountain chain,* sarp *steep,* serpus *head gear,* serd *a setting forth consecutively*

D Ta i̱ra *transcend. excel, go beyond,* i̱rai *anyone who is great,* (cf. 78) *eaves of a house,* araca̱n *king,* ara̱le *post, pillar,* a̱ruvam *greatness, abundance, expanse,* ār *sharpness, pointedness,* ā̱ṟru *become strong, powerful,* ē̱ru *to rise,* ē̱ṟru *lift up, raise*

Y sol- *to fly,* so•tk- *to be strong, able,* sepolah, sepolek *valley field, prairie,* sohci *up, on top, above*

47. HOLD TOGETHER, BE COUPLED (horizontal-bilateral)   *y(e)m-
    (also *y(e)mxʷ(e)-):
    PIE *yem- *hold, hold together, be coupled, conquer,* *yemo- *twin*
    T   yamak *assistant, mate,* yamanmak *be patched on,* yemin *the right hand*
    D Ta yām *we,* yā *to bind, tie;* Kur mīnk̲han̲ā *to close the eyes, eyes to be closed*
    Y   ma•y- *to pass,* micos *male cousin, elder brother,* voc. mit, -me?y *daughter,* meyo•moy *to be pregnant*

48. HOLLOW I, BENT (delative-circumscriptive)   *k(e)w-
    (also *k(e)ws-, *k(e)wfp-):
    PIE *kew-, *kewə-, *kew-k, *kew-b- *bend, bent,* *kūpā *hollow*
    H   kuškuš- *crush*
    T   kaval *hollow pipe,* kavanoz *pot,* kavis *bow, arc, curve,* kavat *wooden bowl,* kubbe *dome, cupola, vault of heaven*
    D Ta kavi̱r *bow one's head, stoop, bend down, be capsized,* kū̱ţu *dome, cupola, nest, etc.,* kūval *hollow, hole, pit;* Ka kappu, kappa *hole in the ground, pit,* kusi *to bend, stoop, etc.*
    Y   kewoy *burden basket,* koyku•?l *hollow rock (lucky)*

49. HOLLOW II, WITH WHICH TO HOLLOW
    (apertive-sublative)   *xʷ(e)r-(fp)
    PIE *orob- *bore, use sharp tool, hollow out*
    T   orak *sickle, reap-hook, harvest,* ordu *army,* orta *middle, center,* ari *bee,* ubur *passage*
    D Ta ka̱ru *stake for impaling criminals,* ka̱rai *pole for propelling boats, elephant goad, etc.,* koru *plough share,* ceru *battle* (c < PD *k-), ceruppu *shoe,* ce̱ru *kill;* Ma kāruka *gnaw, bite*
    Y   kʷar *nail (hammered),* kʷere?wey- *to have a pointed face, to have a sharp tongue,* kʷere?we?y *dugout (canoe)*

50. IN (interpositive)   *(e)n-
    (also *(e)nt(e)r-):
    PIE *en, *n̥ *in,* *enter, *n̥ter *between, into*
    H   anda *inside, between,* andan *inside,* andurza *interior*
    T   enderun *interior, woman's apartment of palace,* ane *pubes, private parts,* an- *understand*
    D Ta nakar *house;* Kui nḍo *at, in, at the place of, in the country of*
    Y   noh(pew-) *to enter,* ?o *in*

51. INDEFINITE, RELATIVE (circumversive-interpositive)   *kʷ(e)n
    (also *kʷene):
    PIE *kʷene *particle of generalization and indefiniteness*

H   kuin *whom*

T   gene *again, moreover, still*

D Ta   aṅku, aṅke (etc.) *there,* äṅku *there, then, thus,* ĭṅku, Ĭṅkan *here,* Ĭṅkanam *in this way*

Y   kʷen, kʷeni *relative adverbs,* kʷelas *he, she, it,* kʷela•kʷ(s) *they,* kus *interrog. pron. and adv.*

52. INJURE, WOUND, PREY UPON (circumscriptive-extenuative) *w(e)l- (also *w(e)lxʷ-, *w(e)lkʷ-):

PIE   *wel-, *wḷ- *injure, wound, plunder, betray, pluck* (Goth. wulwa *rape*), *wḷkʷo-, *lukʷo- *wolf*

H   walh- *strike, attack*

T   veled *bastard, rascal, child, progeny,* ulu *possessors of,* uluma *the howling of dogs,* ulušmak *howl together in packs (wolves,* etc.), ugul-danak *howl (of wind)*

D Ta   vel *to conquer,* val *strong, hard, forceful, skillful,* vali *to be painful,* vāḷ *sword, saw, sharpness,* etc., ūḷai *howl of dog or jackal*

Y   (ʔ)weɫketeg *nail, claw,* (ʔ)wəɫkəʔišneg *wolf*

53. INNER, INWARDNESS (interpositive-horizontal)   *n(e)y

PIE   *eni, *n(e)i *in*

T   in *den* or *lair of wild beast,* yiv *groove,* yiy•ecek *edible*

D Ta   niṟam *bosom, breast, middle place, vital spot,* niṟai *become full, pervade,* nīcuni *go through, pierce*

Y   ni *in*

54. INNER FLOW OR IMPULSE (PASSION, BLOOD, MILK) (interpositive-horizontal)   *n(e)y- (also *n(e)yf-, *n(e)yš-, *n(e)yk-, *n(e)ys-, *n(e)yfkʷ-, *n(e)yft-):

PIE   *nei-, *nei-to *move* or *moved vigorously, excited, passion,* *neid- *flow,* *neid- *abuse, belittle,* *neigʷ- *wash,* *nei-, *neiə-, *nī *lead,* *nēik-, nīk-, nik-, *happen, pounce, begin vigorously*

H   nāi- *lead*

T   yekinmek *make a great effort,* yeğin *active, violent* yeyni, yeğni *light, easy*

D Ta   neyttōr *blood,* ney *butter, grease, fat, ghee,* naṉai *to bud, to appear,* nikaṟ *to happen, occur,* nītu, nīru *to swim in water, overflow*

Y   la•yop *to flow, run (liquids),* la•yok *river*

55. INNER OR BODY SUBSTANCE I (bilateral-circumscriptive)   *m(e)w- (also *m(e)wxʷ-, *m(e)wθ-, *m(e)ws-):

PIE   *meu-, mewə-, mū- *damp, unclean flow, to dirty* also *wash, clean*

H   mauš- *to fall*

T   bev•il, -vli *urine,* bulamak *smear,* bedaub; *dirty, soil* bulaš•tct *contagious, infectious*

D Ta   muṭṭu *pollution, menses,* muka *to draw water,* mūñcu *to lick*

Y   mewec- *to blow (of wind),* muɫs- *to wipe, to lick*

56. INNER OR BODY SUBSTANCE II (circumscriptive-bilateral)   *w(e)m- (also *wemx-, *wemš-):

PIE *wem-, *wemə-, *spit, vomit, (disgust)*

T vehim, vehmi *foreboding, groundless fear, surmise, illusion, delusion,* umma *hope, expectation,* umran *a being in good condition*

D Ma uma *a cough,* ummiṭṭan *difficult breathing, sobbing of children;* Ta umi *to gargle, spit, suck,* umir̠ *to gargle, spit, vomit, emit, be satiated*

Y ʔumeʔwoˑ *river*

57. LEAN, SLOPE, LIE (clinal-horizontal) *č(e)y-
    (also čf(e)y-, *čθ(e)y-?):

PIE *k̂ei-*lie, encampment, cozy,* *k̂oi-to *encampment*

H kitta, kittari *lies*

T čiǧ *avalanche,* čay *stream,*

D Ta toy *to languish, be weary,* etc.

Y keycek *to be tired,* kmoy¹t(kes-) *to lie down, to die*

58. LIKE, ALSO (oppositive-bilateral) *x(e)m-
    (also *x(e)mxʷpe-):

PIE *ambhō(u) dual *both and alike*

H hamenk- *connect, marry*

T âm *general, universal, public, common, ordinary*

D Ta anm *like that,* oppam *comparison, resemblance*

Y *hahpemew *mate, spouse,* *hahpew *wife*

59. LOOSEN, LOOSE PERSON (vertical-extenuative) *š(e)l-
    (also *š(e)l(e)w-, *š(e)l(e)wxʷ-):

PIE *al- *grind,* āl- *err, wander,* *lewə- *cut loose, loosen* (cf. 60)

H alwanzah *bewitch,* alwanzatar *witchcraft*

T salpa *loose, slack,* salivermek *let go, set free,* salak *silly, doltish,* selis *fluent,* silk- *shake (off)*

D Ta allatu *evil, sin,* allavai *sin, evil, uselessness*

Y ɬʔewkʷoh *to be broken,* ɬʔəʔəməc- *to stop* (trans.)

60. LOOSEN WHAT SURROUNDS OR IS SURROUNDED
    (extenuative-circumscriptive) *l(e)w-
    (probably *šl(e)w-, *šl(e)wnxʷ-):

PIE *leu- *cut loose, loosen (wool, hair, fleece),* *lū-no- *cut*

T lâvta *obstetric forceps, doctor* or *midwife,* šev *slope*

D Te valavalana *loose;* Ka nave *to become insignificant in bulk, wear away, become less*

Y ɬʔuɬ(kʷon-) *to explode (of a gun,* etc.)

61. LOVE (expansive-circumscriptive) *s(e)w-
    (also *s(e)wk-, *s(e)wxʷl-):

PIE *s(e)w- *give birth,* suyús *son*

H šawitišt- *nursling, suckling (of animals less than a year old)*

T sevgi *love, affection, compassion,* sevmek *love, like fondle,* sevda *melancholy, spleen, passion, love,* sulh *peace, reconciliation, accord,* sivasmak *adhere,* sivazlamak *stroke, caress*

D Ta avvai *mother,* vekul *to be angry, hate, dislike,* uka *to be glad, pleased, desire,* āvu, āval *desire*

Y skewok(sim)- *to want, wish, love, like*

62. MALE (interpositive-sublative)   **n(e)r-*
    (also **x(e)n(e)r-*, **nert-*):

PIE **ner-(t-)*, **aner- (əner-?) man,* **neryo- manly*

T er *man, manly,* yar *friend, lover,* yarak *penis,* yara *strength*

D Ta erutu *bull, ox, steer*

Y no•lum- *to love, protect,*

63. MARRY, UNITE (horizontal-circumscriptive)   **y(e)w-*

PIE **yew-*, **yu- unite,* **yu-go-m yoke*

T ev *family,* evli *married,* evermek *cause to marry, give in marriage,* yavuk betrothal, yuva *socket*

D Ta iyal *be associated with, agree to, resemble,* etc., iyaivu *union, joining together*

Y (ʔ)we- *to marry*

64. MENTAL ACTIVITY (bilateral-interpositive)   **m(e)n-*
    (also **m(e)ns-*, **m(e)nθ-*):

PIE **men-*, **mena-*: **mnā-*, **mnē- think, be mentally active*

H mene- *sight*

T memnun *pleased, glad*

D Ta en *thought, intention,* anantar *sleep, loss of consciousness, confusion of mind*

Y ʔelek^w *I don't know* cog.?

65. MOTHER, BREASTS (oppositive-bilateral)   **x(e)m-*
    (only as **x(e)mnsy*, **x(e)mnx(e)*:

PIE **am(m)a*, **amĭ mother*

T emcik, emzik *nipple, teat*

D Ta ammā *mother, matron,* aññai *mother,* ammam *woman's breast, food of babies,* amman *goddess,* Tu amani *nipple*

Y hu•k, hu•ksoh *child*

66. MOVE, VIBRATE (oppositive-horizontal)   **x(e)y-*
    (also **x(e)yf-*):

PIE **aig- move mightily, swing, vibrate*

T iyi, eyi *good, well, in good health,* ayar•mak *go astray,*

D Ta ī *fly, bee,* oy *to drag along (as a flood), launch, send forth, give,* ey *to discharge arrows*

Y hewec- *to live, be healthy, get well;* himen *quickly,* himəks- *to hurry*

67. MOVEMENT (apertive-horizontal)   **x^w(e)y-*
    (also **x^w(e)yt-*, **x^w(e)ym-*, **x^w(e)ynθ-*):

PIE **oi-m-*, **i-t-*, **oy-t- movement*

H huinu- *to cause to run,* huiš- *to live*

T eylemek *do,* iyadet *visit* oynak *playful, frisky, mobile,* oyun *game, play,*

D Ta kuni *to dance,* kummi, kommi *to dance with clapping of the hands and singing;* To kwïry- *to move violently threshing about*

Y k^{w}oyc- *to go slowly,* k^{w}əməʔəʔəc- *to drive back, drive away*

68. NEGATION, OTHER (oppositive-extenuative)   *x(e)l-

    PIE   *alyos *other*

    T   el *one other than oneself,* alil *blind*

    D Ta al- *to be not so-and-so,* al *night, darkness,* elā, ellā *Here, you!*

    Y   haɬ, heɬ *hey!,* ho *past time, to*

69. NOTCH, INDENTATION (clinal-interpositive-intromissive)   *č(e)nt-

    PIE   *k̂ent- *pierce*

    T   čentik *notch, defect,* čanta *bag, case, valise, knapsack*

    D Ta taṇtu *stem, stalk, tube*

    Y   tkektkekohs- *to prick* (metathesis?), knetknet *arrowhead, cartridge*

70. OPEN, PENETRATE, DESTROY (apertive-extenuative)   *xʷ(e)l-

    PIE   *ol- *destroy*

    H   hullāi-, hulliya- *fight, overthrow*

    T   uluorta *openly, clearly,* olgun *ripe, mature,* oluk *gutter pipe, groove,* alarga
          open sea, alen•en *publicly, openly*

    D Ta kalappai *plough,* kalai *disperse, be ruined, defeated,* kallu *to dig out, exca-*
          vate, erode, kūlu *fall down, be destroyed, ruined,* keli *to conquer, overcome*

    Y   kla•moks- *to leak,* ko•ɬ *to be open*

71. OPEN UP, RELAX, EXPAND, LOOSEN (expansive-extenuative) *s(e)l-
    (also *s(e)lfč-):

    PIE   *sel- *living space,* *selĝ- *let go, throw, pour out*

    H   šallanu- *to make big,* šalli- *big,* šalleš- *to become big*

    T   salkɨ *hanging, relaxed,* sallamak *swing, rock, shake,* salmak *throw, let go,*
          salpa *loose, slack,* sel *torrent, flood*

    D Ta alar *to blossom, open up, spread,* alaṅku *move, shake, swing, dangle,* iḷaku *to
          become relaxed, get loose, become pliable,* etc. (cf. 59), iḷai *to grow weary*

    Y   sloyk- *to be loose, supple,* siɬ- *to lie, be situated*

72. OVERHANG (delative-extenuative)   *k(e)l-
    (also *k(e)l-x):

    PIE   *kel-, *kl̥-, *kelə- *tower up, high, rise, lift*

    H   kalmara *mountain?*

    T   kelle *head,* kalkmak *rise, start to do,* kaldɨrmak *raise, erect, lift*

    D Ta kala *to appear, come into being,* kiḷar *to rise, ascend, shoot up;* Go kelku *hair*

    Y   klohstoy *to overhang,* keɬpen- *to be thick*

73. POINTING UPWARD (clinal-sublative)   *č(e)r-

    PIE   *ker-, *k̂r̥- *head, horn, peak*

    T   čɨrpɨ *twig,* čardak *hut of brushwood on supports, trellis, pergola, gallows,*
          čarmɨh *cross on which malefactors were crucified*

    D Ta terru *hedge of thorns protecting a passage,* terru-ppal *snagged tooth,* tarai
          to sprout, shoot forth

    Y   keromeca? *sugar pine tree*

74. PRESS TOGETHER (CLOSE EYES) (delative-bilateral)   *k(e)m-
    (also *k(e)mf-):

PIE   *kem- *press together,* *komo- *compressed*
T   kemirmek *gnaw, nibble, corrode,* kambur *hunchbacked,* kör *blind*
D Kurkappnā *cover* or *press gently with hand;* Ta kapōti *blind person*
Y   komtenep *to be blind,* koma *hard*

75. PROTRUDE, GUSH FORTH (projective-horizontal)   *p(e)y-
    (also *p(e)yθ-, *p(e)yxᵂ-):
    PIE   *pey(ə)-, pĭ- *fat, puffed up (milk, juice)*
    T   heyet *shape, form,* heybet *awe, majesty,* heyula *matter (the substance of
          which all is formed)*
    D Ta pituṅku *protrude, bulge, gush out,* pīr *abundant flow, milk flowing from
          woman's breast;*
    Y   mikoy- *to surge,* pemey *grease, fat*

76. REST AGAINST (sublative-bilateral)   *r(e)m-
    (also *r(e)mxᵂ-, *rxᵂ(e)m-):
    PIE   *rem-, *remə-, *rest, support, prop against,* *rom-ti *prop*
    H   armaniya *sicken*
    T   rahat *rest, ease, comfort,* rahim, rahmi *womb,* rampa *incline, loading
          platform,* rehavet *softness, limpness,* zemin *earth, ground*
    D Ta amar *to abide, remain, rest, be tranquil*

77. RISE, ARISE (sublative)   *(e)r-
    (also *erš-, *er(e)w-):
    PIE   *er-, *ere, *erə-(?), *erei-, *ereu- *rise, raise, set in motion,* *r-to *risen* (cf.
          46)
    H   arāi *rise*
    T   ermek *reach, attain,* erišmek *arrive, attain, mature,* artɨrmak *increase,
          augment*
    D Kurargnā *to climb, mount an animal, rise (as sun, moon, stars), lift, raise,
          increase, etc.;* Kol er- *to become, happen;* Ta eṟu *to rise, ascend*
    Y   ʔolonewkᵂ *to float,* ʔolonem- *to carry,* ʔerew(oriš-), ʔelew(oliš-) *to hang*

78. RISEN, RAISED, HIGH RANKING PERSON (vertical-sublative) *š(e)r-
    (also *š(e)rft-, *š(e)rxᵂt-, *s(e)ryxᵂ-):
    PIE   *ardi, ṛdi *point,* *ardh- *stick(?),* aryo- (?) *lord, ruler*
    H   artari *stands*
    T   serefraz *stately, eminent, holding the head high,* serdar *military chief,
          general,* serikâr *supervisor, head of a business,* sarp *steep,* siret *moral quality*
    D Ta ār *sharpness, pointedness, become full, spread over,* araiyaṉ *king, sovereign,
          prince,* iṟa *excel, transcend,* iṟai *anyone who is great;* Ko iṟ- *to have sexual
          intercourse with* (cf. 46)
    Y   ʔo•ʔ *to be, exist, grow (of plants),* roʔoh *to stand, be rooted*

79. SEIZE (WITH MOUTH) (apertive-bilateral)   *xᵂ(e)m-
    (also *xᵂ(e)ms-):
    PIE   *omə- *advance energetically, make fast, admix, harass, injure,* *ₒmīwā
          sorrow

T   am *vulva* umaci̵ *ogre, bogyman*

D Ta   kavvu, kauvu *to seize with the mouth, grasp with eagerness;* Te kamucu *to hold, seize;* Ka gumma *bugbear, devil*

Y   ka•mok *to be disliked,* ka•mes *evil creature, shark,* kem- *to steal*

80. SEVER (sublative-interpositive-apertive-intromissive)   *r(e)nx^w(t)-

    PIE   *rendh- *rip, split*

    H   arhaya(n) *severed* (cf. 27)

    T   zend *forearm, wrist,* zenne *female*

    D Ta   aṟu *be severed, cut, part, break off,* etc.

    Y   leko•(s-), leko•(t-) *stab*

81. SHARE, SUIT, GIVER (vertical-horizontal)   *š(e)y-

    PIE   *ai- *give, share,* *ais- *wish, seek,* *ais- *honor*

    H   -a, -ya, *and, also* cog. ?

    T   seyyan *two equals,* siftah *first stroke of business, first sale of a new commodity*

    D Ta   eytu *approach, reach, obtain, be suitable,* ayyaṉ, aiyaṉ *father, sage, priest, teacher, master,* etc.

    Y   ʔiʔiʔgah, ʔeʔgah *to eat in a group, have a meal*

82. SPLIT (sublative-horizontal)   *r(e)y-

    PIE   *rei- *scratch, split, cut*

    T   zina *bumblebee, fornication*

    D   ari *to cut off, nip off, gnaw, sift, separate, lines in the white of the eye*

    Y   -lin *eye*

83. STRETCH (intromissive-interpositive)   *t(e)n-
    (also *t(e)nft, *tntx^w(e)-):

    PIE   *ten-, tend- *stretch, pull,* tṇ-to *extended*

    T   tenef *guy rope of a tent,* nanik *long nose, snook*

    D Ta   neṭu *long, become tall,* neṭṭam *perpendicularity,* taṉi *be profuse, increase in size;* Pa taṇḍ- *to pull;* Kur tārṇā *to lengthen, draw* or *stretch out*

    Y   to•lekic *stretch a rope as a trap* or *barrier,* noʔmonem- *to endure*

84. STRIVE, INCREASE, DIVINE INTERVENTION (OMEN)
    (vertical-circumscriptive)   *š(e)w-
    (also *š(e)wfk-, *š(e)w(e)ө(y)-):

    PIE   *aw-, *awē, *awēi- *to like,* *awē- *take pains, strain,* *aug- *increase, augment;* Lat. augur *omen,* augos *success of an enterprise granted by the gods*

    H   šagai- *omen,* šakiyah- *to give an omen*

    T   sevk *a driving, urging, inciting,* sevmek *love,* sitayiš *eulogy,* suud *ascent, elevation*

    D Ta   uka *to desire, be pleased* (cf. 61), avalam *suffering, want,* ūkku *to make an effort,* uka *ascend, rise,* vēṭṭam, vēṭṭai *hunting, chase,* vēṇṭu *to want, desire*

    Y   weʔgey *to stretch* intr.

85. SUFFER (circumversive-interpositive-oppositive-intromissive)   *k^w(e)nxt-

    PIE   *k^wenth- *suffer, endure*

    T   kaht *drought, scarcity, famine*

D Ta katuvu *to be troubled, perturbed,* kuṭai *to pain (as the ear, the leg)*
Y kahselop *to feel strange*

86. SWALLOW, NIBBLE, WINK (circumversive-bilateral)   *k^w(e)m-
   PIE *k^wem- *swallow, sip*
   T gem *bit,* gemrenmek *be gnawed* or *nibbled,* gamɨz, gamze *wink*
   D Ta kappu *to gorge, cram in the mouth,* kumaṭṭu *retch, vomit;* Te kamucu *to nibble at, swallow*
   Y miḳolum *to swallow*

87. TAKE I, RELATE TO (bilateral)   *(e)m-
   (also *m(e)s-, *m(e)x^w-, *m(e)xs-):
   PIE *em- *take* (cf. 42)
   H mahhan *like, as*
   T ama *but,* muad•ele *equation,* mah- *hidden, captive, protected, created,* etc., mahsul *product, produce, crop, result*
   D Ta mēvu *to join, reach, desire, love, learn, study, eat, level, make even (as the ground), manifest, assume, abide, dwell, be attached, be united, be fitted* or *joined;* Te ammu *to sell, vend,* mesagu, mesavu *to eat, feed on*
   Y ʔemsi *and,* mosk- *to borrow*

88. TAKE II, SHARE, HAVE (interpositive-bilateral)   *n(e)m-
   (also *n(e)mf(e)-):
   PIE *nem- *share, take*
   H namma *further, then, also*
   T nam *name, quality*
   D Ta nampu *to long for, desire intensely*
   Y negem- *to take,* ʔolonem- *to carry* (for ʔolo- cf. RISE)

89. TAPERED I (clinal-bilateral)   *č(e)m-
   PIE *k̂em- *stick, pole, horn*
   T čam *fir, pine*
   D Ta tappai *bamboo splints for a broken bone, bamboo splits for roofing;* Te tammiya, tamme *lobe of the ear*
   Y kəmətəw *little finger,* kowis *stick;* or < *čel

90. TAPERED II (clinal-extenuative)   č(e)l-
   PIE *k̂el- *slope, incline, bend, thin shaft, pole, rigid stalk*
   T čulik *short piece of tapered wood,* čalɨ *bush, shrub, thicket,* čalɨk *slanting, awry*
   D cf. BEND, FORK
   Y klew *waterfall*

91. TASTE (apertive-bilateral)   *x^w(e)m-
   (also *x^w(e)mr(x^w)-, *x^w(e)mf-):
   PIE *om, *$_o$m- *raw, bitter,* *$_o$m-ro-s, ōmo-s ds., $_o$m-ēd- *eat raw meat*
   T umman *ocean* (salty?), emlah pl. of milh *salt*
   D Ma kamarkka *to have astringent taste;* Ta kampi *bit of a horse's bridle*
   Y ka•mewet *to have a bad taste in one's mouth*

92. TURN, ENTWINE, COVER, SURROUND
(circumscriptive-extenuative)   *w(e)l-
   (also *w(e)l(e)s-):
   PIE   *wel-, wl̥-, *ulē- *turn, wind, waltz*
   T   velense *thick blanket, horse blanket,* uleyk *convulvulus, bindweed,* valih
         bewildered
   D Kol ul *day;* Ta uḷḷi *onion, garlic,* vēli *fence, hedge, wall,* valam-puri *that which
         curls* or *spirals to the right,* valai *net,* vala *to encircle,* veḷḷari *cucumber;* Ka
         vālu, ōlu *to slope, slant*
   Y   woɬkecoy- *to be morning,* woɬkek^w *roof,* woɬur *girl dancer at the Fish Dam
         ceremony*

93. TWIST, MANIPULATE (delative-sublative)   *k(e)r-
   PIE   *kert-, kr̥t- *twist,* *kerd- *handwork*
   T   kerye *iron hoop,* kariš *mix*
   D Ta keruvu *to unite, embrace,* karaṭu *ankle, knot in wood,* kaṟaṅku *to whirl,*
         karuvi *instrument, tool*
   Y   keromekin- *to twist, lock,* keromoh *to turn round (of a wheel,* etc., intr.),
         kelomek *to be twisted,* kelomen- *to turn* (tr.), kelomoh *to turn* (intr.)

94. UPPER BACK BETWEEN SHOULDER BLADES (apertive-bilateral)
                                                                            *x^w(e)m-sf-
   PIE   *om(e)so-s shoulder
   T   omuz *shoulder,*
   D Ta kumpam *upper part of back between shoulders*
   Y   k^woyteme?l *shoulder*

95. WALK, GO, MOVE AGAINST A SURFACE (juxtapositive-horizontal)
                                                                            *θ(e)y-
   PIE   *ei- *go,* *eis- *move quickly, set in motion*
   H   tiya- *to walk,* ;*read*
   T   tay, tayyi *the traversing of a space of time or place, cancellation* or *erasure of
         words,* tez *quick, quickly*
   D Ta tēy *wear away by friction, be effaced, erased, die;* Te diggana, digguna
         suddenly
   Y   teykunow- *to grow together*

96. WANDER, REVOLVE (circumversive-extenuative)   *k^w(e)l-
   (also *k^w(e)lx^w-):
   PIE   *k^wel- *move, wander,* *k^welo-s, k^wek^welo-s *wheel*
   H   kaleliya- *tie, tether*
   T   gel- *come,* galeyan *effervescence* (to boil), kulak *ear, ear-shaped,* kulumbur
         swivel-gun
   D Ta kalai *wander in thought, disperse,* kulunku *to be shaken, agitated,* kuḷacu
         loop, noose, joint of body
   Y   k^womɬec- *to return,* k^wəmɬəyəh(s-) *to turn round* (intr.)

97. WARD OFF, GUARD (circumversive-horizontal)   *k^w(e)y-
   (also *k^wesyrx^w-):

PIE   *k^{w}ei- *watch, show, protect, honor,* *k^{w}ē[i]-ro- *watchful,* protective

H   kuirwana- *protectorate*

T   kuytu *sheltered from the wind,* giriftar *captive,* giy- *wear (clothes)*

D Ka kā, kāyi *to guard, protect, keep, save, tend, watch, keep in check*

Y   ɬkyork^w- *to watch*

98. WEAK, DELICATE (intromissive-sublative)   *t(e)r-
    (also *terw-):

PIE   *ter-, *teru- *tender, weak, slender*

T   ter *moist, green, fresh, juicy,* nerm *soft, mild,* narin *slim, slender, tender, delicate;* tereddi *degeneration, deterioration,* taravet *freshness, juiciness*

D Ta nēr *be soft, grow thin, slenderness,* narunku *be deficient in growth, decay, grow lean, fail,* tār *fall low, degenerate, fail,* etc.; Ma tāṟuka *to sink, decline, become thin*

Y   tel- *to be ill, sick,* telogeɬ *pain,* telogum- *to be in pain,* no•loyk^{w}el- *to be feeble, weak*

99. WEATHER, ENVELOPMENT (oppositive-circumscriptive)   *x(e)w-
    (also *x(e)w(e)s-):

PIE   *aw(e), *awē(i)- *blow, respire,* *aw(e) *moisten, dampen, flow,*

H   heu- *rain*

T   avaz *loud voice,* övünc *boasting*

D   cf. 84

Y   hewomop *to be warm,* hum- to sweat in sweat-house, hewec *to live, be healthy*

100. WEAVE, MOVE, RESTRIBUTE, TURN (oppositive-sublative)   *x(e)r-
    (also *x(e)rs-, *x(e)rx^w-):

PIE   *ar(ə)- *plough*

T   örek *network of a tissue,* örmek *plait, knit,* örü *texture, web,* arka *back, reverse,* arš *forearm, warp of cloth*

D Ta āṟi *circle, ring, wheel,* oṟuku *flow, leak, trickle down,* eṟi *throw, smash, chop,* etc.; Ko a•r- *to do a broad jump*

Y   helomey- *to dance,* holim- *to weave (baskets),* ho•liš *to paddle,* ho•loy- *to be twisted (basket),* ho•lo•t *to wave,* ho•lop̌in *to stir*

101. WEAVE BASKETRY, LONG AND BENT ROUND
    (circumscriptive-horizontal)   *w(e)y-
    (also *w(e)yθ-):

PIE   wey-, wĭ-, weyə- *rotate, bend (often used in basketry, etc.)*

T   via *blood vessel, vein,* yumak *ball (of wool, string,* etc.), yuva *nest,* yuvarlak *round, spherical,* yevm *day,* yiv *groove*

D Ta vēy *to thatch, put on (as a garland),* vai *straw of paddy, grass,* vil *bow,* viṟutu, viṭutu, vīr *aerial root as of the banyan*

Y   weyew *to be woven, to be finished (of baskets)*

102. WITH LIKE ITEMS (emissive-bilateral)   *c(e)m-
    (also *c(e)mx^w-):

PIE   *sem-, *sm- *the same, with,* semo- *anyone,* som- *together*

T   cem *crowd,* cem'an *in all, as a total,* cumhur *the mass of the people*

D Ta cemmai *goodness, agreement*

Y   co•m(-) *to be in a group, be together*

103. WITHDRAWN   *sen-

PIE *seni, senu, (s$_e$ni-), sn-ter- *differentiated, by* or *to oneself, without*

T   sen *thou, you,* sana *to you,* sanih *occurring to the mind,* singin *broken, defeated, routed,* sinir *frontier,* sinmak *break, be routed, be scattered*

D Ta en *thought, intention,* etir *that which is opposite*

Y   son(ow-) *to be, be like, happen, do,* soninep *to feel, think*

Table 6 aligns the reconstructed stems for contrastive purposes. The forms and their identifying (semantically unrefined) glosses are arranged in such a way that initial phememes are identical in horizontal columns and final phememes in vertical columns. This facilitates substitution of phememes in either position and assessment of the underlying semantic ground of that which is kept constant. Table 1 was constructed on the basis of this type of analysis.

This analysis is identical to that carried out in the attempt to discover the meaning of recurrent partials (morphemes) in the words of unanalyzed, or insufficiently analyzed, languages. If one knows the meaning of one part one can determine the meaning of the other by assessing the change as the latter is added (or substituted). A meaning is provided for each of the six resonants in the top, horizontal column. Moving down each column vertically it is easy to see that the upper meaning seems to underlie at least some of the modified forms. It occured to me to make such an analysis when I saw how regularly the same or similar kinds of meaning were associated with a variety of stems in which one phoneme recurred.

To recognize that phoneme and meaningful unit were isomorphic was to open the way toward identification of the cause of the isomorphism. An easy place from which to start this task is an examination of meanings of forms with *m. *Taking* or *relating* involves a bilateral relationship. Allowing for inevitable semantic change that distorts this meaning, or obscures it in some cases, it is not too difficult to see that all of the forms with *m involve some kind of bilaterality: of the fingers or hands in taking or grasping; of mouth and teeth in tasting, chewing, or swallowing; of two opposed surfaces in tapering, pressing together, holding together, crushing, or resting against; of two similar items such as tapered sides, encompassed sides, and likenesses. One such bilaterality characterizes the mouth itself (as /m/ in English *mouth*) another the hand (as /m/ in French *main* or Spanish *mano*).

The sound [m] is articulated by pressing the lips together while air escapes through the nasal passage and the vocal cords are vibrated. The lips provide a bilateral, continued relationship that is repeated in the semantics related to the sound. Articulation and meaning are isomorphic, united through the spatial interaction that the moving parts of the mouth provide.

As we move from front to back of the oral tract we find both similarities and differences between consonants articulated at adjacent points. Sounds involving lip movements have meanings of peripherality, while sounds involving interaction between tongue and teeth or alveolar ridge have internal meanings. In some cases these meanings are more obvious than others, but once we see the direction the analysis is taking, and the kinds of relationships that are involved, it takes only a little imagination coupled with a feeling for system in classification to determine the remainder.

Meanings are all spatial and correlate analogically with relationships effected in the oral tract between articulators. Thus, *p, *f and *m and *w all define external or peripheral space. Articulation of *p involves lip projection, of *f, projection of upper lip and/or upper teeth beyond the lower lip, an abessive or *out from* motion. *m involves lip interaction, an obvious bilaterality. The semantics of *t lead to the supposition that it was articulated as a very dental, perhaps interdental, stop, while *θ articulation involved gentle approximation of tongue to upper teeth. The contrast between *m and *n might better be described as extrapositive vs. interpositive except that for *m the duality of lip contact emphasizes bilaterality. *Movement from* is the feature basic to alveolars and specific directionality of a single surface is basic to palatals, probably stemming initially from the obvious protracted horizontality of the tongue in pronunciation of *y. *w, and its allophone [u], involve lip rounding or total circumscription. It is probable that in enunciation of [w] there was a slight, voiced velarization. In pronunciation of *x^w and *k^w the rounding or circumscription is interrupted by the total or partial velar occlusion, resulting in conceptualization as the opening (opposition) in a partial enclosure for *x^w and a reversion, or circular movement for *k^w. *r, as *sublative,* leads to the supposition that its production involved raising the dorsum, while *š, as *vertical,* suggests tongue retroflection. *k contrasts with *r in emphasizing a downward rather than an upward dorsal movement, appropriate to the delative meaning.

In the framework of PL conceptualization, every event is a spatial event that serves to explain the various meanings that each PL sequence has come to have in present-day languages. Where for us marriage is a social contract, for PL the spatial unification of man and wife is paramount, with interaction of lineal and circular genitalia (Set 63). If spatial relationships are assessed as primary, then metaphoric extension is easily understood. Dream material in psychoanalysis is of exactly this nature; the dream image, like the phonemic image, translates the metaphor into concrete, spatial terms.

As the system evolved through time, the metaphoric analogical process became increasingly abstract until it was not easily recognized that meaning was an analogue of articulation. At the same time repetition of thematic

TABLE 6
(e)R- and C(e)R- stems

INITIAL	FINAL					
	m	n	l	y	r	w
(e)	(e)m—take, relate to	(e)n—in	(e)l—decay, distend	(e)y—go	(e)r—rise	(e)w—enclosure
p		p(e)n—feed	p(e)l—fill, flood	p(e)y—protrude	p(e)r—further	p(e)w—cleanse
t	t(e)m—cut, chew	t(e)n—stretch	t(e)l—delay, stay		t(e)r—weak	t(e)w—care for
c	c(e)m—with like items	c(e)n—complete	c(e)l—extract	c(e)y—flow	c(e)r—flow	c(e)w—centrifugal
č	č(e)m—tapered	č(e)n—notch	č(e)l—tapered	č(e)y—lean, lie	č(e)r—point upward	č(e)w—concave, convex
k	k(e)m—press together	k(e)n—dig, hole	k(e)l—overhang		k(e)r—twist	k(e)w—hollow
kʷ	kʷ(e)m—swallow	kʷ(e)n—indefinite	kʷ(e)l—wander	kʷ(e)y—ward off	kʷ(e)r—form, create	
f		f(e)n—dissolution	f(e)l—emptiness	f(e)y—go forward	f(e)r—height	f(e)n—dress, sur-round

	(e)m	(e)n	(e)l	(e)y	(e)r	(e)w
θ	θ(e)m—grasp		θ(e)l—bend, fork	θ(e)y—walk	θ(e)r—dig up	θ(e)w—emptiness
s		s(e)n—year	s(e)l—open up	s(e)y—go	s(e)r—high	s(e)w—love
š	š(e)m—around both sides		š(e)l—loosen	š(e)y—share	š(e)r—risen	š(e)w—strive
x	x(e)m—like, also	x(e)n—ancestor	x(e)l—beyond, negation	x(e)y—apportion	x(e)r—weave	x(e)w—weather
xʷ	xʷ(e)m—taste, shoulder blades, seize	xʷ(e)n—aperture	xʷ(e)l—open	xʷ(e)y—copulation	xʷ(e)r—hollow	xʷ(e)w—boundary
m		m(e)n—mental	m(e)l—grind, chew		m(e)r—die	m(e)w—inner substance
n	n(e)m—take, share, have			n(e)y—inner (flow)	n(e)r—male	n(e)w—begin
l	l(e)m—crush	l(e)n—bend, sway		l(e)y—dwindle, weaken	l(e)r—deficient ingrowth	l(e)w—loosen around
y	y(e)m—hold together					y(e)w—marry
r	r(e)m—rest against	r(e)n—sever		r(e)y—emit, split		r(e)w—extract
w	w(e)m—inner substance	w(e)n—help	w(e)l—injure, turn	w(e)y—weave basketry		

sequences in association with a metaphor already remote from its spatial origins linked meaning to sequence rather than to individual sound, giving rise to the duality of patterning of separate phonemic and morphemic systems. The linguistic sign had by that time become dissociated from its symbolic beginnings.

Careful analysis of a particular language will allow discovery of some phememic relationships. Such an analysis of PIE would have easily uncovered the meanings of resonants and voiceless stops. Swadesh (1969:51–53) in a discussion of suffixes, and Foster (1969a:145–154), in a discussion of roots, came, quite independently of one another, to the conclusion that phoneme and meaning were to some extent isomorphic on a submorphemic level in Tarascan. Some isomorphisms coincide with those suggested for PL; Swadesh gives *within* as a meaning of /n/; *central* as a meaning of /t-/; *central* or *within* as a meaning of /u/; *extension* as a meaning of /š/; *coming* or *length* as meanings of /y/; while Foster gives *linear extension* for /i/ and /y/; *in place with depth* (i.e., *circumscribed*) for /o/, /u/, and /w/; *separation* for /c/; *protrusion-penetration* for /č/ and /č'/ where meanings of *sharp, cross-wise, zig-zag,* and the like, predominated; *reversal* for both /k/ and /kʷ/; *inner activity,* or *manipulation of space* for /m/; *level* or *change of level* for /p/; *from within* for /r/ (< PL *l); *motion repeated or prolonged* for both /s/ and /š/; *contact* for /t/ (< PL *θ); *pierce, penetrate* for /t'/ (< PL *t). All of this was on the right track, although I had quite forgotten it when analyzing PL, and again reached some of the same, or similar conclusions.

While the first important byproduct of PL reconstruction was the discovery of phememes and their systematization, the second was discovery of bundles of shared isoglosses (innovative features) that could only mean shared early history of certain languages after separation from the others. The only innovation shared by PIE, H and Y is conversion of č to /k/. However, PIE is divided in this, for the eastern languages show /s/ rather than /k/ as a reflex of this phoneme. This is, then, an independent, and not a shared, innovation for the languages concerned.

The branching diagram in Table 7 shows a succession of early separations and the shared features upon which they are based. PIE and H developed voiced stops from some fricative-stop sequences. Development in T of a voiced stop series from *xʷS seems to have come about independently, influenced by prior voicing of xʷ. T loss of *x and *xʷ resulted from their weakening in T, D, and Y earlier stages. It seems necessary to assume that H hardening of *f and *θ also developed independently.

T, D, and Y share innovative developments for resonants. Nasals and liquids split, probably influenced by reduction of earlier FR clusters. Similar splitting of *p and *t occurred in these languages. Another isogloss unites these languages through splitting of *e, probably originally in free variation, after nonvelar fricatives.

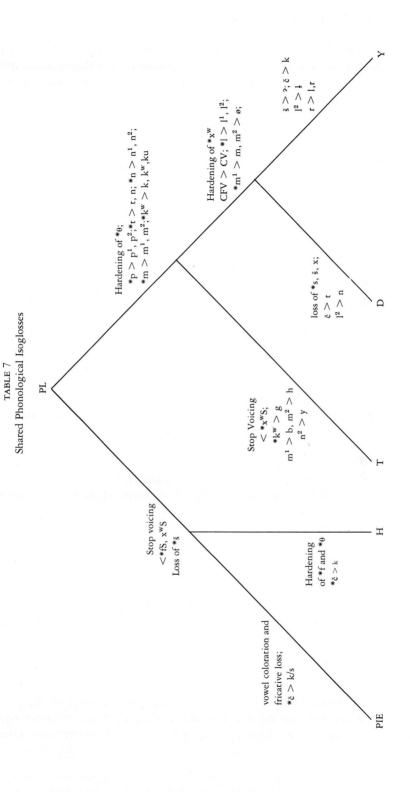

TABLE 7
Shared Phonological Isoglosses

Hardening of *θ;
*p > p¹, p², *t > t, n; *n > n¹, n²,
*m > m¹, m²; *kʷ > k, kʷ, ku

Hardening of *xʷ
CFV > CV; *l > l¹, l²,
*m¹ > m, m² > ø;

š > ʔ; č > k
l² > ł
r > l,r

loss of *s, š, x;
č > t
l² > n

Stop Voicing
< *xʷS;
*kʷ > g
m¹ > b, m² > h
n² > y

Stop voicing
< *fS, xʷS
Loss of *š

Hardening
of *f and *θ
*č > k

vowel coloration and
fricative loss;
*č > k/s

PL

Y

D

T

H

PIE

That T, D, and Y go back to a similar node, attested by the quantity of systematic evidence, does not mean that their relationship is of recent date. The shared innovations will need to be explored for other groups with which these languages have been placed; Algonquian for Yurok, Altaic and Uralic for Turkish.

The relationship between PIE and H does seem to be of more recent date, yet a count of close semantic relationships in this data belies such an hypothesis. There are 12 such close resemblances in PIE and H (Sets 1, 9, 10, 13, 25, 32, 39, 43, 57, 84, 95, 99). T, D, and Y are semantically closer in 18 sets (1, 2, 6, 7, 10, 30, 31, 43, 61, 63, 70, 71, 73, 79, 97, 98, 100), while D and Y show greatest similarity in 11 (11, 24, 25, 54, 55, 56, 57, 68, 74, 82, 85).

The systematic organization of likenesses and differences into an expressive paradigm with duality of spatial reference, hence capable of further metaphoric extension, was undoubtedly the single most important evolutionary advance in biological history. With language as symbol, other aspects of culture were free to develop concomitantly. Explorations of cultural symbolism increasingly show that nonlinguistic cultural manifestations are based on the same cognitive principles as characterize languages. Ethnosemantics, structural anthropology and semiotics are recent theoretical approaches that utilize paradigmatic analysis to achieve significant insights into symbolic organization in culture.

Ethnosemantics analyzes lexical systems into paradigms that are organized by analogy and opposition, but also by a third principle that is not utilized in the patterning of either PL phememes or in modern phonemic systems. This principle is hierarchy, which depends upon inclusion. There is feature hierarchy in PL phememic classification. In lexical hierarchies a dominant term, such as *plant,* shares a semantic feature or features with all subordinate terms for which it may be substituted, as *tree, vine,* or, more specifically, *rose.* Otherwise it is neutral, or featureless, in contrast to the proliferation of features in items of greater specificity, all lower in the hierarchical scale.

Hierarchy in the social sense of physical dominance must have long preceded the emergence of *Homo sapiens;* it characterizes much other animal behavior, as the pecking order of chickens attests. It is not clear to what extent the two types of dominance are cognitively related. The first kind is isolable in other aspects of culture than language; for example, my identity as an American is dominated by representation of the American flag. In this sense my flag can be said to represent me, much as *plant* represents *oak* or *rose.* Further exploration of early lexemic stages should permit discovery of the level at which the principle of hierarchy came into being. It is, of course, always possible that it existed in the lexemic organization of PL. It is still uncertain what constituted a lexeme at the early stage. Sequence equivalences suggest that some syntactic units were already in existence as themes with an initial, simple stem structure followed by a specifying suffix. Fricatives were prefixed, but not resonants or stops.

In concentrating on the paradigm I do not mean to minimize the crucial role of syntax. Paradigms are always virtual, since it is only through syntactic realization that paradigmatic items have their existence. To date, my exploration of various aspects of PL has revealed more of the paradigmatic than of the syntagmatic framework. It seems likely that ordering of units was just as significant to meaning at early stages as at later stages. Rules for organization of phememes into themes with various kinds of suffixal potential awaits analysis. Further reconstruction will provide answers to these and many other crucial questions.

The paradigmatic system that I have abstracted from my reconstruction and presented in Table 1 provides a bridge for understanding the transition from nonlanguage to language. In particular, it shows how the duality of patterning in language came into being. This duality is one in which an open, meaningful sign system is constituted from a limited set of meaningless units; both systems composed of interlocking features, the first articulatory, the other semantic and taxonomically organized.

Table 1 shows a representational system of classification by means of which states and movements in space are translated into spatiosonant, articulatory counterparts. Where movement in nature is unbounded and flowing, with events succeeding one another without spatial delimitation, the representation necessarily segmented experience as it was forced into manipulable units. Classification requires units to classify, and the units of early language were specified by the nature and limitations of the oral tract. Units must be different enough so that they would not be easily confused. Because experience is not naturally, but only conceptually, segmentable, relationships between such segments, to be at all *like* natural experience, must have likenesses as well as differences. Organization of a limited set of units into likenesses and differences is a way of overcoming some semantic problems that arose from the very fact that the set was restricted. Another way that problems could be minimized was by combining the units into strings in such a way that position conditioned meaning. A third way of gaining semantic flexibility was by extending the meanings of the strings metaphorically so that they could be applied to nonspatial relationships that were in some way conceptually similar.

This system shows such an elaborate capacity of the paleolithic mind to organize and classify experience, that it is quite clear that man, by this stage in his biological evolution, had developed far more sophisticated mental powers than have any subhuman species today. Since we must assume that until this organization took place man did not have language, all of this mental activity was necessarily covert. It seems probable then that other symbolic activities were being carried out that were equally covert and not transmitted through language. Burial practises, for example, which have always been assumed by anthropologists to require language before they could occur, seem not to rely on language at all but only on some observable, common manner of symbolizing. The genesis of language was the genesis of only one manner of symbolizing among many, all of which were designed primarily to represent, and only secondarily to communicate.

As the communicational uses of language began to expand, language itself began to change in significant ways. Greater practice and speed in articulating long strings of symbolic units tended to change the ways in which sounds were pronounced. Inclusion of metaphoric extensions with the original meanings of groupings, coupled with the tendency of suffixes and prefixes to become inalienably attached to stems, began to erase the similarity between sound and meaning that was clearly apparent in the early stages. Duality of patterning was born. Despite the changes, the organizational structure retained all of the earlier modalities of analogy, opposition, and taxonomy. These were renewed as they were lost in order to keep the system intact. The linguistic sign seemed unmotivated and arbitrary to Saussure (1959:67–70) because there was no detectable relationship between sound and meaning. In reality, even after this relationship had become obscured, the patterning and interrelationship of features, both of sound and meaning, continued to be so firm that we can hardly consider today's linguistic sign to be arbitrary. If it is not arbitrary it is clearly symbolic.

Speculations on and discussions of language origin consistently fail to take into account the complexity of feature interrelationship characteristic of all human languages. Such feature patterning seems to be totally lacking in the communicational repertory of any subhuman species, including apes that have been taught by humans to communicate. Failure to recognize this seems curious since many of the authors and discussants are linguists and as such are thoroughly familiar with the covert complexities of human language.

Systematic linguistic complexity, even at the very early period represented in Table 1, depends upon a covert manipulation of analogy and opposition to connect and differentiate units and groups of units. Wherever units are combined into groups, and these units into greater groups, a taxonomy exists, whether or not the nodes are overtly distinguished.

Analogy, opposition, and taxonomic organization are the three classificatory principles that underlie every human symbolic system, including language. Nonhuman animals, and especially primates, are capable of analogical classification, but there is nothing to date in the ethological record that indicates that they are capable of utilizing either opposition or taxonomy and putting it to the service of communicational representation. If studies of animal behavior were to focus on underlying mechanisms of sign behavior typical of *both* human and nonhuman species, a clearer picture might emerge of just how the human animal got that way.

Notes

1. Foster, in 1969b and 1970, but particularly the latter, provided a germinal presentation of this hypothesis, since modified and refined.

2. For example, Chomsky writes, ". . . the structure of a phonological system is of very little interest as a formal object; there is nothing of significance to be said, from a formal point of view, about a set of forty odd elements cross-classified in terms of eight or ten features" (1968:65).

3. A preliminary isoglossic exploration is found in Foster, 1971, distorted to some extent by failure to eliminate borrowings into Turkish from Arabic.

4. Other Dravidian abbreviations which occur sporadically in the data are: Te (Telugu), Tu (Tulu), Ka (Kannaḍa), Pa (Parji), Kol (Kolami), Ko (Kota), Kui (Kui), Kur (Kuruẖh), Ma (Malayalam), Malt (Malto). The lexical material on which the reconstructions are based derives from the following sources: for PIE, Pokorny, 1959; for H, Friedrich, 1967, and, occasionally, Sturtevant, 1964, or Puhvel, 1965 or 1966; for T, Hony, 1957; for Y, Robins, 1958; for D, Burrow and Emeneau, 1961.

5. Phonological abbreviations used in the discussion are: C = consonant, V = vowel or vocalic allophone, S = stop, F = fricative, R = resonant. The canonical stem under consideration, then, is either R, a single, vocalic resonant, or CR, consonant plus resonant. C may represent any consonantal phoneme, either as consonantal or vocalic allophone for phonemes with both.

6. Benveniste (1962) demonstrated that /e/ vocalism does not occur in the IE root if it occurs in a suffix, and vice versa. This hypothesis has yet to be explored for PL.

7. Pokorny (1959:316–317) equates this form, I believe erroneously, with PIE *enek-reach, attain.

8. The H form is included because it has been considered to be cognate with the PIE form. This presents difficulties that must be resolved by consideration of such reflexes as T ağiz mouth; a problem that lies outside the stem canon convention considered here. Luw = Luwian, a Hittite-related Anatolian dialect. This word is undoubtedly derived from fleece = (sheep's) covering.

9. The meaning of this etymon does not correspond in any obvious way to the bilateral-sublative interpretation but raises interesting questions about early death practice or belief.

References Cited

Benveniste, E.
 1962 Origines de la formation des noms en Indo-Européen. Paris: Adrien-Maisonneuve, Librairie d'Amérique et d'Orient.

Berlin, Brent, and Paul Kay
 1969 Basic Color Terms: Their Universality and Evolution. Berkeley: University of California Press.

Burrow, T., and M. B. Emeneau
 1961 A Dravidian Etymological Dictionary. Oxford: Oxford University Press.

Chomsky, Noam
    1957 Syntactic Structures. The Hague: Mouton.

    1968 Language and Mind. New York: Harcourt, Brace and World.

De Laguna, Grace Andrus
    1963 Speech: Its Function and Development. Bloomington: Indiana University Press.

Foster, Mary LeCron
    1969a The Tarascan Language. University of California Publications in Linguistics 56. Berkeley: University of California Press.

    1969b Ten postulates for primordial language reconstruction. Paper presented at the 68th annual meeting of the American Anthropological Association in New Orleans.

    1970 Explorations in semantic phylogeny. Paper presented at the 69th annual meeting of the American Anthropological Association in San Diego, California.

    1971 American Indian and Old World languages: a model for reconstruction. Paper presented at the 70th annual meeting of the American Anthropological Association in New York.

Friedrich, Johannes
    1967 Hethitisches Elementarbuch II. Heidelberg: Carl Winter-Universitätsverlag.

Haas, Mary R.
    1958 Algonkian-Ritwan: the end of a controversy. International Journal of American Linguistics XXIV, 3:159–173.

Hony, H. C.
    1957 A Turkish-English Dictionary. Oxford: Clarendon Press.

Kuryłowicz, Jerzy
    1927 ə indo-européen et ḫ hittite. In Symbolae Grammaticae in honorem Ioannis Rozwadowski, Vol. 1. Pp. 196–218. Cracow.

Lancaster, Jane B.
    1968 On the evolution of tool-using behavior. American Anthropologist 70:56–66.

Mounin, Georges, and discussants
    1976 Language, communication, chimpanzees. Current Anthropology 17, 1:1–21.

Pei, Mario, and Frank Gaynor
    1969 A Dictionary of Linguistics. Totowa, New Jersey: Littlefield, Adams.

Pokorny, Julius
    1959 Indogermanisches Etymologisches Wörterbuch. 2 Vols. Bern and München: Francke Verlag.

Polomé, Edgar
    1965 The laryngeal theory so far. In Evidence for Laryngeals. Werner Winter, ed. Pp. 9–78. The Hague: Mouton.

Puhvel, Jaan
    1965 Evidence in Anatolian. In Evidence for Laryngeals. Werner Winter, ed. Pp. 79–92. The Hague: Mouton.

1966 Dialectal aspects of the Anatolian branch of Indo-European. *In* Ancient Indo-European Dialects. Henrik Birnbaum and Jaan Puhvel, eds. Pp. 235–247. Berkeley: University of California Press.

Robins, R. H.
1958 The Yurok Language: Grammar, Texts, Lexicon. P. 14. University of California Publications in Linguistics 15. Berkeley: University of California Press.

Sapir, Edward
1921 Language: An Introduction to the Study of Speech. P. 23. New York: Harcourt, Brace (Harvest).

Saussure, Ferdinand de
1879 Memoire sur le système primitif des voyelles dans les langues indo-européenes. Paris: Vieweg.

1959 Course in General Linguistics. Pp. 67–70. New York: Philosophical Library (first edition 1916).

Sturtevant, Edgar H.
1964 A Comparative Grammar of the Hittite Language (second printing, revised edition, 1951. Copyright 1933, the Linguistic Society of America). New Haven: Yale University Press.

Swadesh, Morris
1969 Elementos del Tarasco Antiguo. Mexico: Universidad Nacional Autonoma de Mexico.

1971 The Origin and Diversification of Language. Chicago: Aldine Atherton.

Wescott, Roger W., ed.
1974 Language Origins. Silver Spring, Maryland: Linstok Press.

Dr. Beach and experimental animals.

Frank A. Beach

Human Sexuality and Evolution

Beach shows that maleness and femaleness are two patterns of biologically determined sex differences that incline the individual's development in the masculine or the feminine direction. These biological differences are defined and perpetuated in every human society, but the precise forms may differ widely. Sexuality is a product of human evolution, and the sexuality of humans differs in fundamental ways from that of other primates.

INTRODUCTION

I am going to begin my essay on a personal note to explain why I chose this particular topic. My doctoral thesis dealt with the effects of brain lesions on maternal behavior in female rats. That study appeared in the *Journal of Comparative Psychology* in 1937, and practically all of my research during the ensuing 36 years has been concerned with the neural, hormonal, and experiential control of sexual behavior in animals. As I conducted and published my experiments over that long span of time I slowly came to consider myself a minor authority on sexual behavior. Pride goeth before a fall, and my fall came last year when I initiated an undergraduate course entitled "Human Sexuality."

Teaching that course taught me that although students were quite willing to listen to lectures about mating behavior in animals and biological factors involved in human sex life, they considered these subjects ancillary to the major issues. They were disappointed by my failure to come to grips with basic psychological and sociological problems of sexuality. I on my part was disturbed by their inability to understand, or my inability to demonstrate to them, the essential unity of the so-called biological and psychological approaches to our subject, and the potential value of viewing human sexuality in evolutionary perspective.

This essay is the child of my frustration. It represents a first approximation of a theory aimed at reconciling the different needs and points of view that gave rise to what I perceived as the students' dissatisfaction and to my own feelings of uncertainty and inadequacy.

SEX VERSUS SEXUALITY

The arguments I am going to summarize are neither original nor profound but even so they eluded me until three orienting conclusions gradually became clear. These conclusions are as follows: (1) All laymen and nearly all psychologists and psychiatrists implicitly and unconsciously distinguish between male-female and masculine-feminine as separate, dyadic categories. (2) The fundamental issues underlying most treatments of sexuality center upon masculinity and femininity and have relatively little to do with maleness and femaleness as such. (3) Nearly all problems traditionally dealt with in theories of sexuality are much more relevant to the behavior of men and women than to that of males and females of other species.

When I had gotten this far in my thinking I began to suspect that I might achieve further insights by starting from a working hypothesis that sexuality is a product of the evolution of *Homo sapiens*. If this were true, however, if sexuality were a product of evolution, it almost surely would be an evolutionary *emergent*. That is to say, it would represent a new level or form of organization not apparent in previous stages of evolution—one involving saltatory change from preexisting levels and not predictable from a knowledge of the characteristics or qualities of those levels. It is my present feeling that human sexuality is about as closely related to the mating behavior of other species as human language is related to animal communication, and that relationship is distant indeed.

The synoptic survey of facts and theories that follows is organized in terms of ontogenetic history of the individual and phylogenetic history of the species. Individual development is subdivided into the epigenesis of maleness and femaleness on the one hand and of masculinity and femininity on the other.

The term *epigenesis* is chosen deliberately to emphasize the fact that development of structure and of behavior involves a process of gradual diversification and differentiation in which new traits, characters, or attributes actually develop out of initially undifferentiated ones. Ontogeny is not merely a process of accretion in which new features are superimposed on those present in the preceding stages. It is a progression from one level of organization to the next in the course of which new and quite different characteristics, relationships, and functions appear at successive levels. As a matter of fact, epigenesis might be thought of as *emergent ontogenesis.*

EPIGENESIS OF MALE AND FEMALE

In every sexually reproducing species a fundamental difference between males and females is permanently and irreversibly established at the moment of fertilization. In all mammals the haploid nuclei of the paternal and maternal gametes fuse, and the diploid nucleus of the zygote contains either two X chromosomes or one X and one Y. This means that the nucleus of every somatic cell in the individual produced from that zygote will contain either two Xs or an X and a Y chromosome that are direct descendants of the original sex chromosomes.

As epigenesis progresses through embryonic and fetal stages the process of sexual differentiation results in the appearance of more and more differences between males and females. For example, at an early point in embryonic life the gonads of both sexes contain cortical and medullary components. In XX embryos the cortex gives rise to an ovary, and in XY embryos the medulla differentiates into a testis. A new sex difference has emerged as a consequence of the original difference in chromosomal balance.

Hormones secreted by the embryonic testis have a directive effect upon the subsequent growth and development of accessory male and female sex structures. Embryos of both sexes possess primordial tissues that have the capacity to develop into male or female sex accessories, but in the genetic male, testicular hormone stimulates the Wolffian ducts to differentiate into the epididymis, vas deferens, and seminal vesicle, while the Mullerian ducts fail to develop. In female embryos there is no testis hormone and the embryonic ovary is inactive. Since the Wolffian derivatives are not stimulated and the Mullerian system is not suppressed, the latter differentiates into the fallopian tubes, uterus, and median vagina.

After these changes have taken place male and female organisms are distinguishable in terms of differences in the sex chromosomes, structure of the gonads, secretion or lack of secretion of gonadal hormone, and anatomy of the sex accessories. Figuratively speaking, the gap between male and female is steadily widening as development proceeds.

In anticipation of conclusions to be developed later it is important to emphasize here that these successively appearing differentiating characters will persist throughout the entire existence of the individual, and that they will contribute to development of additional sexual dimorphisms at much later stages of ontogeny. For example, the uterus, which differentiates in the female embryo, eventually will be essential to menstruation, and beginning at menarche the occurrence of menstrual cycles will indirectly affect development of some of the psychological characteristics that differentiate adolescent girls from boys of the same age.

Sex differences mentioned thus far are established in the embryo. Additional ones appear during the fetal stage of development. Compare, for example, an XY fetus bearing testes, epididymis, and vas deferens with an XX fetus possessing ovaries, tubes, and uterus. The gonads of the XY fetus secrete testosterone (which is not the same hormone produced by the embryonic testis), while those of the XX fetus are endocrinologically inactive. Testosterone contributes to the development of still further divergences between males and females. In the male, primordial tissues of the urogenital sinus differentiate to form a penis and a scrotum. In the female the same anlagen give rise to a vagina and a clitoris. There are now at least five separate characters distinguishing males from females.

Sex differences in reproductive anatomy are striking and easily observed, but there are others more difficult to discern and equally important in terms of their effects upon the eventual development of sex-related behavior patterns. I refer to differences in the male and female brain that are established during prenatal life. The existence of such differences in animals has been established by experiments involving exposure of unborn females to stimulation by testicular hormone and observing their sexually dimorphic behavior as adults. Numerous studies have shown that prenatal androgen treatment decreases female behavior and increases the incidence of mating responses typical of the male (Gorski 1971). If male animals are deprived of the testosterone normally secreted by their own testes, their sexually dimorphic behavior in adulthood is more like that of females than is the behavior of control males, and they show clear deficiencies in behavior characteristic of their own sex (Beach 1971).

Behavioral changes induced by altering the endocrine environment during development provide at best presumptive evidence for sex differences in the brain, but recent discoveries prove the existence of neuroanatomical differences in brains of male and female rats (Field and Raisman 1973). The differences, which are present at birth, can only be detected by electromicroscopy and are restricted to the one small area in the hypothalamus. Nevertheless, their significance is difficult to exaggerate for the region in which they occur is known to be involved in mediation of sexually dimorphic mating behavior. Furthermore, brains of genetic females can be caused to develop like those of males by exposing the females to androgen in utero, and brains of males castrated at birth develop postnatally to resemble those of normal females.

All of the sex differences mentioned to this point have been qualitative rather than quantitative. The variables involved are discontinuous rather than continuous but it is important to emphasize the existence of quantitative differences between males and females that represent the operation of continuous variables.

With very rare exceptions every newborn child can be assigned unequivocally to one of two populations designated as male or female. The variables in terms of which these populations differ are numerous, and for most measures there is marked interindividual variability within both populations. In fact, for the majority of sex-related traits, differences between male and female can only be expressed in terms of means and standard deviations, and there is appreciable overlap between the distributions of the two populations.

This does not make such sex differences unimportant, but it is important to recognize their relative nature. For example, at birth female infants are, *on the average,* ½ to 1 cm shorter, and 300 g lighter than males. Newborn females are, *on the average,* four to six weeks more advanced than males in skeletal maturation. Infant females are, *on the average,* less muscular, less active, slightly more sensitive to pain, and more irritable than males. When the inventory of congenital sex differences is examined *in toto,* it appears that the only characters for which males and females are qualitatively dichotomous are those in which dichotomy is essential for reproductive function, for example, testis versus ovary, uterus versus epididymis, or penis and scrotum versus clitoris and vagina.

Many differences between human males and females have been established by the time of birth and additional differences continue to emerge through the childhood years, but sex differentiation is markedly accelerated during the period of adolescence. Some of the more obvious changes occurring at this time involve development of the secondary sex characters. In girls the externally observable changes include growth of the breasts, development of female habitus with broadening of the hips, and onset of menstruation. Analogous changes in boys involve change in pitch and timbre of the voice, growth of the penis, occurrence of seminal ejaculation, gradual appearance of facial hair, and eventual beard development.

Less obvious but equally significant changes occur in various physiological functions. Figure 1 shows that in adolescence the muscular strength of boys increases to a much greater degree than that of girls, and Figure 2 demonstrates similar differences in blood pressure and blood hemoglobin. Marked increase in the secretion of testicular hormone illustrated in Figure

127

3 contributes to these and other changes marking male adolescence. Appearance of female secondary sex characters is linked to the beginning of cyclic secretion of estrogen and progesterone by the ovaries. Taken in combination, the dimorphic secondary sex characters reflect a new stage of divergence between males and females. The final degree of separation will be achieved even later when females undergo pregnancy, parturition and lactation.

The essential features of the epigenesis of male and female which I have so sketchily summarized seem to me to be the following. Starting from an original difference in chromosomal balance, the individual passes through a succession of steps or stages of development at each one of which new differences emerge. Some are mutually exclusive whereas others are expressions of continuous variables that produce quantitative differences. Sexual differentiation is best viewed as a developmental progression from one level to the next. At each level new organization appears that was not discernible at earlier levels. The essential process is not one of addition or accretion but of organic growth and differentiation. I believe that the development of sexuality is also epigenetic and is in various ways analogous to the development of sex.

EPIGENESIS OF MASCULINE AND FEMININE

Development of most of the basic differences between male and female occur between fertilization and birth. In the space of nine months the foundation is laid for the eventual expression of all male-female dichotomies that are present in the neonate or will arise in the course of the individual's lifetime. In contrast, differences between masculine and feminine do not even begin to develop until the second or third year after birth, and their full maturation occupies a span of 15 to 20 years. When we transfer our attention from male and female to masculine and feminine, we are shifting from the analysis of differences in sex to the analysis of differences in sexuality.

The concept of sexuality involves two components, gender role and gender identity. Two gender roles are defined by every human society. They comprise a composite of all those behavioral traits, attitudes, and emotional characteristics that combine to define and differentiate masculine and feminine members of that society. Some dimensions of sexuality are very much the same in all societies, but others exhibit extreme intersocietal variability. The evolution of gender roles will be considered in the final section of this essay. At this point I wish to analyze the epigenesis of masculine and feminine, or the development of gender identity.

Gender identity pertains to each person's feelings and convictions regarding his or her own sexual makeup. More specifically, it refers to the individual's concept of his or her own masculinity or femininity. This is an exceedingly important dimension of the more general self-concept, which in turn plays a vital role in overall psychological development. If a society is to function effectively it must establish and maintain viable gender roles. If individual members of a society are to function effectively they must develop integrated and stable gender identities.

I have already suggested that the development of the individual's gender identity progresses through successive stages, taking new form or organization as new levels emerge. The principal stages recognized and utilized by every society are infancy, childhood, adolescence and maturity. As defined by most societies, gender roles in early childhood are relatively simple, that is, they involve few variables, and become increasingly complex and more finely differentiated at each successive stage of postnatal ontogeny.

As already noted, all societies lay the foundation for development of gender identity by assigning every new infant to one of two gender roles. Now, despite the fact that the majority of male-female differences are quantitative rather than qualitative and involve extensive overlap between the sexes, all societies divide their newborn members into two discontinuous classes, and this exceedingly influential decision is based upon one dichotomous character, the genital anatomy. The significance of this fact is that sex role ascription at birth will have powerful and continuing influence on development of the individual's gender identity. The implicit and intuitive definition of male and female as discontinuous and mutually exclusive categories leads inevitably to a bipolar, unidimensional concept of masculine and feminine. This concept violates the biological facts, so to speak, and in some societies, including our own, it can do mischief by encouraging compartmentalized and opposing stereotypes of masculinity and femininity that create serious problems not only for the society but for many individuals as well.

Sex role ascription has no significance for the newborn infant, but for the society it predetermines which one of two programs of training will be employed in the sexualization of that individual over the next 15 or 20 years. In contrast to sex role assignment, the actual start of gender identity development coincides more or less with the onset of verbal learning. Disregarding the simple conditioning of which neonates and even fetuses may be capable, we can date the emergence of gender identity from the start of language acquisition.

Development of gender identity depends upon two categories of learning. The child must learn the distinguishing characteristics of the gender role of both sexes and must also learn to behave in a manner consonant with his or her assigned gender role. The growing child starts to learn about sex differences as soon as he begins to use and understand language. As a first step each male learns to identify himself as a boy and not a girl, although at first this means no more than saying, "I am Johnny."

By three years of age or earlier most middle class American children can correctly state their own sex, and at four years they can identify the sex of dolls of various ages, relying principally upon differences in hair style and clothing. One group of three-year-olds was required to assemble cutouts of male and female figures separated into three vertical sections consisting of the head, the trunk, and the area below the trunk. When the figures were distinctively and differentially attired most children performed successfully, but when the cutouts were nude many children had difficulty identifying the sex of the section bearing the genitals. Heads were correctly matched with trunks, but errors with respect to the below-the-trunk section were made by 88 percent of three-year-olds, 69 percent of four-year-olds and 31 percent of six-year-olds (Katcher 1955). Other studies have shown that American children do not achieve a clear understanding of genital differences until they are five to seven years old, even when there has been express parental instruction.

An important concept that young children master with difficulty is that of gender constancy. In one study children from four to eight years old were shown pictures of a boy and a girl and were asked whether the girl could become a boy if she so desired. Most four-year-olds were sure that she could do so if she would cut her hair, change to boys' clothes, and play boys' games. Only children six years and older were confident that even if she made these changes the child in the picture would still be a girl (Kohlberg 1966).

Certain aspects of adult gender roles are learned fairly early in life. These include not only such simple items as sex-related differences in dress but also various aspects of adult daily routine. For example, in most middle class families five-year-old children know that fathers go to work in the daytime whereas mothers stay at home or go shopping. They may also know that older brothers play football while older sisters engage in less strenuous forms of recreation.

The essential point at issue is the fact that a growing child's concept of male and female gender roles is not easily and quickly acquired. Instead, it is built up gradually over a number of years as a result of observation and instruction. Even so, this learned discrimination by itself is not sufficient to insure development of the individual's gender identity. It is not enough for a little boy to learn how men and women behave or how his society expects and desires him to behave. He must learn to govern his behavior so that it conforms to those expectations. Furthermore, he must, in most respects at least, *enjoy* fulfilling his assigned gender role. Successful development of gender identity involves three requirements which are: (1) to know how to behave in accordance with gender role, (2) to want to behave in the appropriate manner, and (3) to be able to execute the appropriate behavioral patterns. Achieving concordance of all three variables is what gender identity is all about. Incongruities between any two can interfere with the epigenesis of masculinity or femininity.

Fortunately, but not at all fortuitously, in all societies many of the major dimensions of childhood gender roles are correlated with basic male-female differences characteristic of the human species. Such differences are reflected in the fact that in all societies that have been carefully studied, boys are more active, venturesome, self-assertive, and physically aggressive than girls. Girls tend to be more sedentary, more dependent, and less aggressive. These sex-related differences are manifest in early childhood. For example, one review of observations of American children indicates that, ". . . in virtually every relevant study of preschool and school-age children, aggressive behavior has been found to be more frequent among boys than girls. Boys also show more negative attention-seeking and antisocial behavior than girls. Even in their fantasies . . . boys indicate a greater preoccupation with aggressive themes" (Mussen, Conger, and Kagan 1969:504).

Descriptions of behavior of children in numerous other societies reveal a marked intercultural commonality of such early sex differences. For instance, in his account of one Melanesian society, Davenport observes that boys of six years and younger characteristically engage in a great deal of rough-and-tumble play, gang fighting, and other forms of vigorous social interaction. Little girls in contrast play quietly in small groups or alone (Davenport 1965).

Societies in general appear to have intuitively recognized certain male-female differences, formalized them to some extent, and incorporated them in the stereotyped gender roles for the two sexes. The way in which initial sex differences in aggression are modified and molded in our society is indicated in the following quotation.

> Aggressive behavior is an accepted component of traditional masculine behavior (i.e., sex-typed behavior) but not of feminine behavior. Aggression in girls typically meets with more punishment than it does in boys, and the role-models young girls choose are less likely to be overtly aggressive. For these reasons young girls who are aggressive will gradually learn to inhibit aggressive manifestations, while boys have more freedom to express their hostile feelings and will continue to manifest aggressive behaviors. [Mussen, Conger, and Kagan 1969:341].

The authors of this quotation note elsewhere that American boys may be encouraged to behave aggressively in certain situations and condemned when they fail to do so.

In the Melanesian society studied by Davenport strict gender role training begins as soon as children can walk. Little boys are allowed to run freely all over the village, forming age-stratified gangs. They range about, "fighting amongst themselves and getting into mischief, always under the close but permissive surveillance of adults". Little girls, in contrast, "are virtually never separated from their mothers or older sisters while the latter are engaged in household and garden work before sunup to past sundown" (Davenport 1965:195).

These examples could be multiplied almost indefinitely but they will suffice to illustrate one hypothesis I wish to suggest concerning differences between males and females and between cultural definitions of masculine and feminine. Using aggression as an example, the hypothesis begins with the reasonably conservative assumption that prior to and in the absence of any social training, punishment, or reinforcement, human males and females differ in their tendencies to interact actively, vigorously, and aggressively with other individuals. It is further assumed that the strength of these tendencies varies considerably among individual members of both sexes, although the average value for the male population significantly exceeds the average for the female population. In other words, the situation is that which obtains with respect to sex differences in all continuously varying traits. The picture is one of two distribution curves with reliably different means but a considerable degree of overlap.

Starting from this base, society imposes a distinction in gender roles that embodies the original male-female difference but eventually exaggerates it by reducing or eliminating the area of bisexual overlap. Little boys who tend to be less aggressive than the average for the male population are encouraged to display more aggressivity. Little girls who show more aggression than the average for the female population are punished and encouraged to behave less aggressively.

To the extent that social rewards and punishments can modify aggressive behavior, the net result of the foregoing process will be to produce a kind of regression toward the mean in both male and female populations that in turn increases the apparent bimodality of the two distributions. Society has, so to speak, used the gender role differences that it defines in the first place as a wedge to drive between the two populations and thus artificially widen the distance between them. Society did not create the original difference, but society did seize upon that difference and exploit it for purposes that will be discussed in connection with the evolution of human sexuality.

I suggest that this hypothesis is germane to many gender role differences that have wide cross-cultural generality. Consider for example sex differences in the games children play. Many of them reflect or exploit underlying and unlearned male-female differences in strength, agility, physical endurance, and other biological variables. In numerous societies around the world boys throw at targets, run races, catch balls, climb trees, and in general prefer activities for which most males are physically better suited than most females by virtue of their superior control over large muscle groups and better eye-hand coordination in gross movements. Girls in the same societies tend to choose games that demand less energy expenditure and capitalize upon female superiority in fine muscle coordination and rapid and precise small movements.

Concordance between the childhood activities that societies designate as masculine and as feminine and congenital sex differences that render these activities selectively suited to the average boy's and the average girl's abilities and interests, has the result that most members of both sexes tend to enjoy the actualization of their sex roles, practice them frequently, and become proficient in their performance.

All of this constitutes and contributes to the development and growth of a masculine or feminine gender identity, but the repertoire of human male-female differences is not confined to physical capacities or sensory-motor coordinations.

Observations of children in different societies reveal that boys are more curious than girls about their environment and are in general more likely to ask questions starting with "why" or "how." Boys tend to be more analytical than girls and independent in solving problems (Smith 1933). Boys are also more persistent. In one experiment children were given two puzzles, one of which they were able to solve, while failing the other. When later allowed to return to one puzzle, boys tended to choose the one they had failed whereas more girls preferred to repeat their earlier success (Crandall and Rabson 1960).

Among American children of school age more boys than girls display a preference for the study of science and mathematics while more girls than boys prefer literature. In IQ tests adolescent girls tend to excel on verbal measures, whereas boys of similar age are superior on quantitative and spatial problems (Kagan 1970).

Earlier in this essay I mentioned the existence of a neuroanatomical sex difference in the brains of newborn rats. The demonstration of emotional and intellectual differences between boys and girls raises the age old question as to possible differences between male and female brains in our own species. Although most scientists today would probably withhold judgment on this issue it is amusing to recall that no such uncertainty plagued the minds of many authorities less than a century ago. The following excerpt is taken from a medical treatise entitled *Cerebral Hyperaemia* written by William A. Hammond, M.D. in 1895.

> Certainly, my experience goes to establish the fact that the study of mathematics is bad for the average young woman's mind. I have repeatedly had instances of cerebral hyperaemia under my charge occurring in young ladies of from

fifteen to seventeen years of age, in whom it was directly induced by the study of calculus, spherical trigonometry, and civil engineering. I have now the care of a young lady, sixteen years of age, in whom the disease came on rapidly, in consequence of long-continued and close application to the solution of a mathematical problem. But so long as there are ambitious women who want their sex to study all the subjects men do, I suppose civil engineering will be responsible for many hyperaemia brains in young girls. [Hammond 1895:71–72]

Leaving aside the validity of Dr. Hammond's theories concerning the etiology of cerebral hyperaemia in adolescent girls, it is a fact that gender role specifications and the components of gender identity undergo several important changes beginning at the time of puberty. Some of these are correlated with the new physiological and anatomical differences that characterize the onset of adolescence and serve further to distinguish males from females.

Representative sex differences have already been illustrated in Figure 1, which reveals the increasing disparity between boys and girls with respect to muscular strength. Figure 2 shows differences in oxygen carrying capacity of the blood (red cells) and capability of the circulatory system to deliver fresh blood to the sensory and motor mechanisms for behavior (systolic blood pressure). These differences are functionally related to some of the gender role assignments discussed in the final section of this essay.

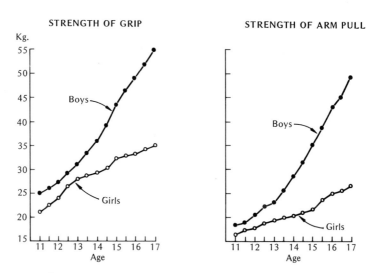

FIGURE 1
Sex differences in muscular strength at different ages. (Based on data presented in Tanner 1962.)

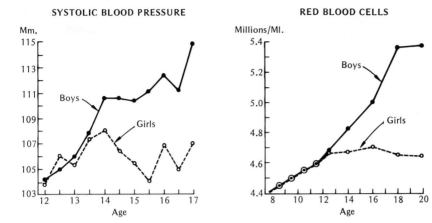

FIGURE 2
Sex differences in blood pressure and erythrocyte count at different ages. (Based on data presented in Tanner 1962.)

They are also differences that at the time of their development exert indirect effects upon the epigenesis of gender identity. For example, the adolescent boy's concept of his own masculinity is affected by changes in his physical capacities, which are apparent to himself and to his associates.

Behavioral changes associated with puberty and with the increase in testosterone shown in Figure 3, include occurrence of the first seminal ejaculation at a median age of 13.8 years and beginning of masturbation at about 12 years. Nocturnal emissions appear at this time and are most frequent between the ages of 12 and 16 years (Ramsey 1943). Events of this nature serve to focus the boy's attention upon genital aspects of sexuality and contribute to development of a masculine gender identity.

In many societies the occurrence of menarche is associated with additional qualifications of the feminine gender role, and at the same time the beginning of menstruation contributes new dimensions to the adolescent girl's gender identity. It is certain that changes in society's attitudes and expectations play an important role in modifications of adolescent gender identity, but account must also be taken of the major physiological changes that occur at the same time. For example, the relatively sudden increase in secretion of ovarian hormones is not without effect upon behavioral tendencies. Psychologists have compared the social behavior and interest patterns of pre- and postmenarchial girls of the same chronological age. Post-menarchial girls were more interested in daydreaming, personal adornment, and display of personal and social activities with the opposite sex (Stone and Barker 1939).

The last item is especially significant. The emergence of strong

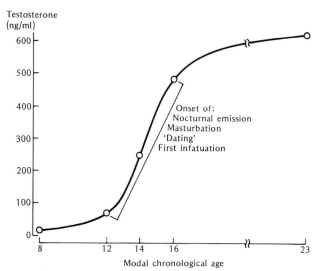

FIGURE 3
Levels of plasma testosterone in human males at different ages
(August, Grumbach, and Kaplan 1972). Data referring to behavior
taken from Ramsey (1943), and Kephart (1973).

heterosexual attraction is a very common element in the development of
adolescent gender identity for both sexes. The broadening of interpersonal
associations to include heterosexual activities is characteristic of this stage of
psychosexual ontogenesis. Many societies indirectly acknowledge the fundamental change by restructuring gender roles between late childhood and
early adolescence. Rules governing sex-related behavior of adolescents may
range from active encouragement of dyadic heterosexual liaisons on the one
hand to sweeping prohibitions against social contact between teen-age
males and females on the other (Ford and Beach 1952). Both extremes
reflect implicit recognition of the fact that important and powerful changes
in sexuality are to be expected during puberty.

In our own and many other societies one striking manifestation of
pubertal changes in gender identity is referred to as *falling in love*. One study
of more than one thousand American men and women (Kephart 1973)
indicated that for most people three important components of *romantic love*
are (1) strong emotional attachment toward a person of the opposite sex, (2)
tendency to think of this person in an idealized manner, and (3) pronounced
physical attraction, "the fulfillment of which is reckoned in terms of
touch."

Eighty-four percent of the respondents recognized romantic love in
terms of their own experience, and of these 90 percent differentiated be-

tween love and infatuation, although they felt the discrimination could only be made in retrospect. The average age for beginning dating was 13 years in both sexes. For males the first infatuation had occurred on the average at 13.5 years and the first love affair at 17.5. Corresponding means for females were 13 and 17 years.

It seems to me we need not look far for the biological correlates of this new and emergent level of gender identity development. No one would assume that there is a one-to-one relationship between the physiological alterations occurring at puberty and the psychologically complex processes of falling in love. On the other hand it can scarcely be pure chance nor yet completely due to social conditioning that timing of an initial, sustained upsurge of heterosexual attraction in adolescents of both sexes coincides so closely with the marked rise in secretion of gonadal hormones.

The foregoing discussion of the epigenesis of masculinity and femininity is obviously incomplete and important features of psychosexual differentiation have necessarily been omitted. The essential points I have tried to make are that gender identity is a fundamental variable in the broader concept of human sexuality, that it develops in the individual over a long period of time under the influence of socially structured programming based upon each society's gender roles, and that some of its dimorphic components, though certainly not all of them, are closely associated with and develop out of biological differences between the human male and female.

PHYLOGENESIS OF SEXUALITY

The final section of this essay is concerned with interrelationships between sexuality and human evolution. In the first two sections I outlined the main features of sexual differentiation and differentiation of one component of sexuality, namely gender identity. I tried to show that certain salient differences between masculinity and femininity can be directly or indirectly linked to correlated differences in maleness and femaleness. The second component of sexuality is gender role, and gender identity is defined in terms of gender role. Masculine and feminine gender roles differ in detail from one society to the next but there is, nevertheless, considerable cross-cultural similarity.

The questions to which the following discussion is addressed are the following. Where did gender roles come from? How does it happen that basic concepts of masculinity and femininity are fundamentally so similar from one culture to the next? What are the biological functions and social values of separate gender roles for males and females?

I wish to emphasize at the outset that what I am now going to present is neither a theory to explain the evolution of *Homo sapiens* nor an attempt to account for all aspects of human sexuality. It constitutes at most a working hypothesis, a preliminary attempt to interpret masculine and feminine gender roles as both products and determinants of human evolution.

Evolutionary change depends upon genetic variation and natural selection, and the genetic unit upon which selection operates is not the individual but the gene pool as represented by members of an interbreeding population. Individuals do not evolve; populations do. If it is to evolve, a population must survive, and the key to survival is reproductive success—reproductive success of the population rather than of the individual.

The traits or characteristics specific to any species include not only anatomical and physiological features but also emotional tendencies, behavioral capacities, and intellectual capabilities. Characters favored and preserved by natural selection tend to be those that make the greatest contribution to survival and reproduction of the interbreeding population. Evolution involves continual and reciprocal interaction between the group genotype and the results of its phenotypic expression, and this includes a *feedback effect* of behavior upon the genotype. Behavioral patterns or tendencies that maintain or increase the probability for survival and reproduction of the group will be preserved and perpetuated. Counteradaptive patterns of behavior will, in the main, fail the test of natural selection and will not be incorporated into the species genotype.

According to this point of view the total behavioral repertoire of any species can be regarded as a mechanism or aggregation of mechanisms reflecting that species' successful solution of the problems of survival and reproduction. Species-specific behavior is adaptive for the simple reason that it has been repeatedly and continuously tested and molded by the forces of natural selection. Different species have different kinds of behavior for the same reason that they have different types of teeth, limbs, digestive systems, and skeletons; because these anatomical, physiological, and behavioral characters have contributed to and have been preserved in the evolution of the species.

One behavioral characteristic man shares with all other primates is that of being a highly social animal, and organized patterns of group living can contribute in various and significant ways to both survival and reproduction. Human society is properly classified as an adaptive mechanism, and this implies that our species-specific patterns of interindividual behavior are what they are because they have withstood the test of natural selection and have contributed to the evolution of our species.

If it is assumed that the human form of social organization has been both a means and a product of man's evolution, then it should be possible to analyze the adaptive function of those forms of behavior that characterize human society, and one ubiquitous characteristic of all human cultures is the dichotomous categorization of their members on the basis of sex, and the assignment of different gender roles to males and to females. The question therefore arises as to how sexuality and society are interrelated and in what ways sexuality may have affected and been affected by the structuring, preservation, and perpetuation of human society.

Any serious attempt to answer such questions soon reveals that sexuality is an essential element in the very warp of human social life. It is true but trivial that without *sex* the human species could not survive for more than one generation. It is immeasurably more significant to recognize that without *sexuality* society in its human form could neither survive in the future nor have evolved in the past.

Before attempting to document these sweeping conclusions I must enter one caveat. In speaking of human society I do not refer to present-day America, or Europe, or even to the Egyptian dynastic societies of 3000 B.C.E. My analysis deals with human social organization as it existed some eight or ten thousand years ago, prior to the domestication of plants and animals and prior to the invention of agriculture. Reasons for this restriction are reflected in the following quotation.

> Even 6000 years ago large parts of the world's population were nonagricultural, and the entire evolution of man from the earliest populations of *Homo erectus* to existing races took place during the period in which man was a hunter. The common factors that dominated human evolution and produced *Homo sapiens* were preagricultural. Agricultural ways of life have dominated less than 1 percent of human history and there is no evidence of major biological changes during that period of time. . . . The origin of all common characteristics must be sought in preagricultural times. [Washburn and Lancaster 1968:213–214]

Preagricultural man was not only a hunter, he was also a gatherer. In fact, even when game was plentiful at least 70 to 80 percent of his diet probably consisted of plant material. In addition to being a hunter-and-gatherer, early man was a social animal; and as his prehuman, primate progenitors evolved into group living forms, their evolution had to include development of certain basic controls over interindividual behavior. For social groups to survive and reproduce, intragroup conflict and competition must be modulated, and in addition the social group must serve adaptive functions above and beyond those provided by the behavior of its members acting independently.

It is generally agreed that as a hunting-and-gathering species preagricultural man probably lived in small groups, rarely exceeding 20–50 individuals (Goldschmidt 1959). Paleontological evidence suggests that early man killed, butchered, and brought to the group's living place a variety of animals, including species sufficiently large and dangerous so that they could only have been overcome by concerted and coordinated action by a number of individuals (Washburn and Lancaster 1968). Coordination of hunting necessitated not only momentary cooperation but planning in advance, and also some form of leadership. It seems reasonable to assume, therefore, that the social structuring of primitive man's life involved foresight, cooperation, and at least temporary hierarchical distribution and acceptance of authority.

Once a kill had been made and meat brought to the living site there had to be some form of food sharing, a type of behavior that is rare in any living primate except between a mother and her offspring. Sharing was a necessity to insure the survival of nonhunting members of the group who were in all likelihood chiefly females, infants, and subadults. Although females probably did not hunt regularly, they were far from being an economic liability to the social group. It was primarily the females who gathered seeds, berries, roots, and other vegetable foods. These also were collected at the living place and shared with other members of the group.

Several theorists have emphasized the importance of sex-differences in reproductive function as factors influencing the evolution of human social systems (Zuckerman 1932; Chance and Mead 1953) and the foregoing reconstruction of prehistoric man's society points to an additional dimension in the foundations for differentiation of gender roles. Such differentiation was not based solely upon differences in sexual and parental functions but also upon sexually dimorphic economic specialization that was coupled with sex differences in physical attributes. Women are hypothesized to have been gatherers rather than hunters because successful hunting depended upon masculine strength and endurance, and perhaps even certain emotional tendencies such as less fearfulness and greater willingness to venture far from the safety of the home base. The more sedentary role of females was also directly associated with the restrictive effects of pregnancy, the necessity of remaining with the young during the lactation period, and possibly a stronger tendency toward nurturant behavior in general.

It is a general rule of organic evolution that a genetic mutation, which must of course occur in one individual, will gradually spread and invade the group genotype provided it does not reduce the potential of the bearers for survival and reproduction and it has no deleterious effects upon perpetuation of the gene pool. Furthermore, among such mutations the ones most likely to be preserved by natural selection are those that most directly contribute to survival and reproduction of the interbreeding group.

It is here suggested that when man evolved into a hunter and gatherer he exploited, so to speak, preexisting sex differences that are easily observed in nonhuman primates today. Males are stronger, more aggressive, more dominant, and more powerful than females. They have greater physical endurance, are less fearful, more investigatory, and hence, more likely to venture into and explore unfamiliar environments. Of course, nonpregnant and nonlactating females possess all of the same traits and capabilities but on the average they are less strongly developed in the female population than in the male. Both females and males possess other characteristics that are just as important to group survival and are more highly developed in the feminine sex.

As the hunting-and-gathering way of life developed in the protohuman primate, the aforementioned sex differences assumed new functional significance. Male genotypes that were above average in promoting those characteristics specifically related to effective performance of the hunter role were especially adaptive from the point of view of group survival. Within the female population, natural selection favored perpetuation and dissemination of those gene patterns that contributed most to behavior consonant with nonhunting, with gathering, and with remaining near the home base. The basic notion suggested here is that, regardless of sex differences in reproductive function, human evolution could easily have involved selection pressures that contributed to gradual and progressive genetic divergence between males and females. These selection pressures then increased the ease and proficiency with which men and women adopted and performed those sex-related gender roles that had adaptive value for the reproducing population.

Reconstructions of the life of prehistoric man are not products of armchair theorizing. They are based in part upon such archaeological evidence as large collections of elephant, rhinoceros, and baboon bones with flint arrows embedded in them and signs of prolonged human occupancy nearby. They are also founded upon intensive study of the few surviving contemporary societies that subsist on the basis of a hunting-and-gathering economy. Sahlins (1959), who has compared twelve such societies, believes that their social organization closely resembles that of preagricultural man.

All of these peoples live in relatively open groups that maintain friendly relations with each other. All such societies practice division of labor by sex; and the sharing of food and other items is universal. As a matter of fact, such sharing is regarded by Sahlins as a *sine qua non* of the human condition.

> Food sharing is an outstanding functional criterion of man. . . . Among all hunters and gatherers there is a constant give and take of vital goods through hospitality and gift exchange. Everywhere, generosity is a great social virtue. Also general is the custom of pooling large game among the entire band, either as a matter of course, or in times of scarcity. [Sahlins 1959:66]

Perhaps the most central aspects of human sexuality and gender role are those pertaining to interindividual relationships within the family, and human family structure probably evolved *pari passu* with evolution of the species. Heterosexual pair bonding for reproductive purposes is well known in fishes, birds, and some types of mammals, but it is rare in primates. Very few species of monkeys form stable family groups and no examples have been described for man's nearest living relatives, the chimpanzee and gorilla.

Seen against this background of anthropoid behavior, *Homo sapiens* stands out as a striking exception. There is not today and probably has never been a human society that lacked the essential elements of family structure.

In his generalized description of present-day hunting-and-gathering peoples and their social structure, Goldschmidt (1959) stresses the fact that they live mostly in small bands, and that each band is divided into families or hearth groups, consisting of a marital couple and their immature and unmarried children. Sahlin's painstaking analysis of 12 primitive societies reveals that their social organization always includes formation of domestic units through marriage. He adds the following important note, "In the domestic economy of the family there is constant reciprocity and pooling of resources" (1959:66).

Evolution of the nuclear family may well have constituted an essential feature of man's phylogenetic history. In fact, differentiation of the family unit within the large social group probably represents a diagnostic character of *Homo sapiens*. If this is true—if the family evolved while man was evolving as a hunting-and-gathering, group-living, tool-making, language-using, culture-transmitting primate—what factors were responsible? Family structure is not essential for individual survival within the group nor for the production of young. It is not even necessary for rearing of young by their mothers or for the protection of young adult male members of the society. All of these necessities are met in modern primate groups that include no family structure.

One unique aspect of human evolution is reflected in Sahlin's reference to the domestic economy of the family and his observation that this includes, "constant reciprocity and pooling of resources (1959:66)." For a theory of human sexuality that necessitates selective advantage conferred by emergence of the family, Sahlin's statement contains a clue that has been followed up by Washburn and Lancaster (1968).

They point out that in nonhuman primates mother-young subgroups are stable and long lasting, and in addition, the social relations within the entire group are ordered by positive affectional ties and by the strength of personal dominance. Both sorts of socializing influences persist in human society in which mother-young bonds are intensified by prolongation of the dependency period. In human society, however, a new element is added. This is the element of economic reciprocity, and economic reciprocity creates an entirely new set of interpersonal bonds. In preagricultural society men hunted, women gathered, and subsequently they shared each other's spoils. The conclusion is as follows, "According to this view the human family is the result of the reciprocity of hunting, the addition of a male to the mother-plus-young social group of the monkeys and apes" (Washburn and Lancaster 1968:224).

Economic reciprocity is only one factor that might help explain evolution of the human family. Others include the ubiquitous tendency and even necessity to form strong and enduring emotional bonds with particular individuals within the social group, and, in all likelihood, a specific need for heterosexual bonding that includes, but also exceeds, the need for sexual gratification.

We have now seen that in the course of human evolution the gradual differentiation of masculine and feminine gender roles served various adaptive functions including those of reproductive and economic significance, but still another function of such behavioral dimorphism was related to the neotony that characterizes man's ontogenetic development. The young of our species are much slower than those of any other primate in attaining a state of physical independence. What is at least equally important is the fact that attainment of complete social or behavioral maturity is an even more gradual process. This is true in large measure because developing the capacities and techniques for adult roles in human society necessitates many years of learning and experience.

The young of preagricultural man had to acquire specific skills involved in the *survival behavior* of his species, but he also had to learn how to *be human.* This included learning many patterns of interindividual behavior essential to the preservation and perpetuation of his society. Central to such social maturation was learning to fulfill the respective gender roles which over hundreds of thousands of years had slowly evolved, differentiated, and been preserved because they conferred adaptive values upon the human social group. Learning to meet the demands of the masculine gender role would have been facilitated by possession of emotional traits and intellectual capacities differing at least in degree from those contributing selectively to rapid and efficient learning of the feminine gender role. One result could be gradual selection favoring genetic linkages in which female reproductive functions would be tied to a bias in interests and capacities fitting the individual for maximal efficiency in learning and living the feminine gender role. Precisely the opposite pattern of selection may have operated on the XY genotype in an evolving population.

Regardless of the validity of such speculations it is clear that evolution of the family not only expanded the functional significance of gender roles but also provided an especially appropriate social environment in which young males and females could learn what their gender roles were and how to meet the associated requirements.

One other possible function of family structure, which has received a great deal of attention by certain theorists (Zuckerman 1932; Chance and Mead 1953), is the opportunity it provides for regular and frequent heterosexual intercourse. A cornerstone in such theories is that fact that unlike females of other primate species most women do not undergo regular behavioral cycles involving periods of sexual responsiveness and attractiveness which alternate with phases of complete indifference or even opposition to sexual congress.

These hypotheses can be identified generically by their acronym, ERV. They are theories of the *Ever-Ready Vagina*. Their basic postulates are as follows: (1) Since she lacks a clear-cut behavioral estrous cycle, the human female is constantly receptive sexually. (2) The human male is constantly potent and needful of sexual relief through copulation. (3) Evolution of the nuclear family occurred because the female's unvarying receptivity gratified the male's perpetual sexual demands.

The ERV theory is logically and biologically untenable for a number of reasons. (1) No human female is constantly receptive. (Any male who entertains this illusion must be a very old man with a short memory, or a very young man due for a bitter disappointment.) (2) Human males do not live in a constant state of sexual readiness or need. (3) Even if women were receptive at all times this would represent only one of two essential features of behavioral estrus as seen in other primates.

Estrus in nonhuman mammals is a phase marked not only by willingness to receive the male but also by changes in behavior and sensory qualities that make the female sexually exciting and attractive to the male. (Males do not become aroused by and rarely attempt to copulate with nonestrous females.) The ERV theory's assumption of constant receptivity in our species is twice damned because it fails to explain the implicit assumption that females exhibit constant sexual attractiveness and capacity to arouse males. According to this theory, the constantly receptive female should be constantly exciting, constantly capable of arousing males sexually. What are the characteristics that make the human female attractive to males even when estrogen levels are low and she is not about to ovulate? (4) Female animals in estrus show by their behavior that coitus at that time has *rewarding* or *reinforcing* consequences. When they are in heat female rats will exert extra energy, perform learned tasks, or endure electric shock in order to achieve copulation with a male. Anestrous females will not do these things. What is the reward or reinforcement for copulation in the human female? Many explanations have been offered in other contexts but the entire problem is overlooked by proponents of the ERV theory.

An even more serious weakness of the ERV theory is its almost exclusive concentration upon sex and neglect of sexuality. A second theory that shares the same defect will be mentioned only in passing. It is one that can be titled acronymically the URP, or *Unready Penis* theory. Proposed in the late 19th century by a French anatomist, this interpretation of human evolution is based upon the accurate observation that among all primates, man is the only species lacking a penile bone or baculum.

The theory holds that primate species in which males possess an os penis devoted inordinate amounts of time and energy to sexual activity, and that by virtue of his deficient genital anatomy the protohuman male was left with leisure time which he employed to invent tools, develop language, and create culture. Except for its curiosity value, the URP theory has few virtues. Even its major premise is wrong. Instead of being less active sexually than other primates, human males and females probably engage in copulation at a much higher overall frequency than animals of any other anthropoid species.

One point stressed by the ERV theory is correct and extremely important. I refer to the absence of behavioral estrus in the human female. The evolution of *Homo sapiens* involved no basic changes in most aspects of the female's reproductive cycle. Her monthly rhythm of ovigenesis, ovulation, estrogen and progesterone secretion, uterine stimulation and menstrual bleeding follows the basic primate pattern. There are, however, two conspicuous deviations from this pattern and are both behavioral. (1) The preovulatory rise in estrogen levels is not accompanied by a reliable and pronounced increase in the woman's desire for and responsiveness to sexual stimulation, and (2) the same change in estrogen secretion does not suddenly render her sexually more attractive to males.

Nonhuman primate females actively solicit and engage in copulation when estrogen from the maturing ovarian follicle is present in high concentrations and ovulation is imminent. Males do not show marked erotic arousal in the presence of anestrous females, but when the female has been exposed to endogenous or exogenous estrogen she evokes strong sexual excitement in the male, which is manifest in his prolonged inspection of her vagina, prompt erection, vigorous mounting, intromission, and eventual ejaculation.

This complex synchronization of hormonal states in the female with behavioral tendencies in both males and females automatically results in copulations occurring predominantly at the time the female is fertile. The adaptive value of such synchrony is so obvious that its disappearance during human evolution is perplexing, and it is apparent that compensatory modifications had to occur that would ensure that a sufficient number of matings would take place when females were about to ovulate and thus could be impregnated. One obvious solution would involve an increase in the overall frequency of copulation across successive stages of the female's cycle, but this change would in turn necessitate alteration of both female and male sexual behavior.

It would be necessary (1) that men find women sexually attractive and copulation with them satisfying even when little or no estrogen was being secreted by the ovaries, and (2) that at these same times women experience the desire for sexual stimulation and capacity for gratification in response to it.

This essay is not the place for a detailed consideration of human sexual motivation but several points relevant to evolutionary changes may be mentioned briefly. Unlike female monkeys and apes, women about to ovulate do not exude sexually exciting odors or develop changes in size or coloration of the perianal skin. Women do, however, possess permanently dimorphic secondary sex characters that are lacking in other primates and are considered erotically exciting in all human cultures. The breast of the human female is unique among primates both in terms of its anatomical location and its prominence in nulliparous and nonpregnant individuals. Furthermore, cross-cultural comparisons identify it as a source of sexual arousal for males and a locus of erotic sensation for females (Ford and Beach 1952).

Another species difference relates to certain characteristics of the human vagina. An important factor contributing to the frequency of intercourse is the amount of pleasure and satisfaction derived during the act of copulation, and this in turn depends in part on the quality and intensity of genital sensations. If coitus is physically difficult or painful its reinforcement value is decreased or eliminated. In the experimental laboratory male monkeys may occasionally attempt copulation with ovariectomized females, but the male's performance under such conditions is desultory and when it occurs intromission does not result in ejaculation. When the female is injected with estrogen the male's sexual interest quickens, his mating performance becomes much more vigorous, and insertion leads to ejaculation.

Part of the change in the male's attitude is due to changes in the female's odor and general demeanor, but nonejaculatory copulation may also be influenced by the lack of vaginal secretion in the absence of estrogenic stimulation. The human vagina is stimulated by estrogen, but it also has the capacity for self-lubrication by means of a nonsecretory process that Masters and Johnson (1966) have termed *sweating*. Vaginal sweating occurs under conditions of sexual excitement and is not dependent upon ovarian hormones, which may help to account in part for the fact that human males and females find intercourse rewarding at all stages of the female cycle.

Although women do not automatically reveal chemically or by innately organized patterns of movement their state of sexual readiness, the human capacity for symbolic behavior provides a more than adequate replacement for such signals. By the use of language and ritualized facial and body gestures (some of which may represent cultural universals (Eibl-Eibesfeldt 1970), women in all societies are capable of conveying to male members of the same societies their sexual desire and availability.

Even though female mammals are sexually receptive during physiological estrus, they may not respond immediately and positively to a male's investigations or attempts to mate. In all species the male is provided for such eventualities with a repertoire of ancillary behavior that tends to stimulate a temporarily unresponsive or resistant estrous female and arouse her sexually to the point of permitting or even soliciting coitus. The same *courtship* or precopulatory stimulation will not elicit receptive behavior on the part of the female when she has not been exposed to ovarian hormones.

Although human females definitely are not continuously sexually receptive, they are continuously copulable and their *sexual arousability* does not depend on ovarian hormones. This relaxation of endocrine control contributes to the occurrence of coitus at any stage of the menstrual cycle.

As far as the basis for sexual satisfaction in human males and females is concerned, this is a subject too complex and extensive to be covered in one essay. It is, however, interesting to note that sources of reward for the copulating female have not been dealt with by evolutionary theories in general.

It appears to have been taken for granted that the primary reinforcement for sexual behavior in male mammals of all species is the experience of orgasm which is assumed to accompany ejaculation. Such an assumption needs testing when applied to other species and is inadequate to account *in toto* for the human male's sexual activity, but the linkage of positively reinforcing orgasm to the ejaculatory reflex necessary for emission of sperm has obvious adaptive connotations. Since the male's ejaculation is necessary for insemination of the female, association of the act of copulation with sensations that tend to reinforce that act and increase the probability of its recurrence would indirectly contribute to perpetuation of the species.

How are we to account for the female's participation in the procreative act? Since survival of our species can hardly have rested upon rape or prostitution, any evolutionary theory needs to consider the problem of sexual reinforcement in females. It is not necessary that women enjoy intercourse in order to become pregnant, but if pregnancy is to occur often enough to ensure continuity of the species, it is important that copulation be rewarding for both sexes. No simple explanation could possibly account for feminine participation in coital relations but there is one source of reinforcement that can be directly tied to changes occurring during human evolution. This element is the phenomenon of the female orgasm.

Thirty-five years of research on sexual behavior in animals have led me to the tentative conclusion that female orgasm may be a biological rarity—a human invention, so to speak. Orgasm is by definition a subjectively recognized emotional response and therefore never accessible to direct observation. We cannot know whether a female chinchilla or chimpanzee experiences orgasm, but neither can we tell whether a woman does so except on the basis of her report of her own feelings. The same statement applies to the human male, but we nevertheless tend to assume that observed ejaculation is a reliable indicator of orgasm in other individuals and in males of other species.

The observable responses of women to genital stimulation also include specific reactions that can be correlated with the report of orgasmic sensations and there is substantiating evidence in the form of electromyographic recordings from the vagina and uterus revealing a close temporal relationship between spasmodic muscular contractions and the woman's subjective experience of orgasm (Masters and Johnson 1966).

From the comparative or cross-species point of view a significant fact is that as far as the male is concerned ejaculation and its behavioral sequelae, including temporary impotence and sexual disinterest, can be identified in the mating behavior of many animal species, whereas behavior indicating the occurrence of sexual climax in copulating females is extremely rare. There may be one or two species that should be excepted, and individual members of still other species may not conform to the general rule, but the weight of available evidence favors the theory that female orgasm is a characteristic essentially restricted to our own species.

If the capacity for female orgasm is an evolutionary product, this capacity could be functionally related to several anatomical and behavioral characteristics that separate human beings from other primates. The first of these is correlated with bipedalism and the associated necessity for upright posture, two characteristics that mark man as unique among the primates. Structural changes occurring in connection with the evolution of bipedal progression and permanent vertical posture included a forward tilting of the human pelvis, and this change in turn resulted in repositioning the female genitalia so that the vaginal aperture was moved to a more anterior or ventral position. One result of this change was increased accessibility of the vagina to penile insertion during ventro-ventro copulation.

The species-typical mating position for all terrestrial mammals including monkeys and apes involves copulation a posteriori whereas the coital pattern predominant in all human societies is some variant of the face-to-face position (Ford and Beach 1952). This may be the so-called missionary position in which the man mounts the woman from above, or the relationship may be reversed, or intercourse may take place with both partners seated, or lying on their sides. Many other positions are known and employed, but ventro-ventro coitus is the most frequent type in all cultures.

The relevance of this fact to the occurrence of female orgasm is that all types of face-to-face intercourse involve apposition or juxtaposition of the pubic symphyses of the two sexes, and continuous or rhythmic pubic contact tends to result in secondary or indirect stimulation of the clitoris. In many cultures assumption of the superior position by the woman is recognized as one method of maximizing her sexual pleasure and satisfaction and this is true partly because she can adjust her position and movement to control the degree and frequency of clitoral excitation.

Despite learned disputations regarding the psychological significance of vaginal versus clitoral orgasm, research has shown that the vaginal and uterine contractions which provide the proprioceptive sensations underlying orgasms are identical regardless of the source of peripheral stimulation (Masters and Johnson 1966). It is equally well established that for nearly all women the most reliable source of stimulation leading to orgasm is mechanical agitation of the clitoris.

The evolutionary argument therefore runs as follows. Human copulation is most frequently carried out a anteriori because of anterior placement of the vaginal orifice which has been indirectly affected by man's upright posture. Face-to-face intercourse results in clitoral stimulation. Stimulation of the clitoris increases the probability of feminine orgasm. Since capacity for orgasm is independent of woman's current hormonal status, the possibility of its occurrence throughout her cycle tends to reinforce and increase the frequency with which she desires and accepts intercourse.

The ERV theory holds that evolution of the human family was related to the disappearance of behavioral estrus in the female. It postulates that absence of fluctuations between receptivity and nonreceptivity permitted sexual relations at all stages of the female's cycle and thus contributed to formation and maintenance of interpersonal bonds of an enduring nature.

An alternative interpretation could be that the loss of behavioral estrus and consequent disappearance of synchronization between mating and feminine fertility necessitated an increase in copulatory frequency so that intercourse would occur by chance at the time when it could result in fertilization. Development of the family may thus be viewed as an adaptive change tending to encourage frequency and regularity of heterosexual coitus. It is a reasonable assumption that existence of the family with the associated intensification and prolongation of interpersonal bonds and dependencies would promote more frequent intercourse than would be likely to occur in the absence of family structure.

This is probable for several reasons. For one thing, familiarity between family members would allow simplification and increased efficiency of the signalling systems employed by both parties to indicate sexual desire and readiness. For another, the family setting would reduce disruptive interference.

In primate societies that lack heterosexual pair bonding, copulation between adult males and females often is prevented or interrupted by harassment by other individuals. Such activities are common in the *anubis* baboon as described by DeVore (1965). A striking contrast is seen in Kummer's account of *hamadryas* baboons whose society consists of family groups comprising one adult male, several females and their young. In this species males copulate only with their own females and other individuals do not interfere (Kummer 1971).

In similar fashion the social setting of the human family tends to eliminate interference with intercourse resulting from social competition, jealousy, and other negative influences. Finally, family structure would be likely to result in couples copulating more often simply because of longer periods of proximity.

Since copulation tends to result in mutual physical gratification, and since simple learning promotes association of positive values with perceived sources of reward, it follows that intrafamilial copulation would provide one source of reinforcement of emotional bonding between males and females. In other words, sexual behavior would reinforce family structure, and family structure would reinforce sexual behavior.

CONCLUDING REMARKS

This essay has been concerned with the ontogenetic development of sexual attributes in the individual and the phylogenetic development of sexuality in the course of human evolution.

From the instant of conception the single-celled zygote incorporates a genetic program that, in the normal course of ontogeny, will result in the selective differentiation of either male or female characteristics ranging from anatomical features of the reproductive system to a repertoire of sex-related behavioral tendencies. Maleness and femaleness connote two patterns of interrelated, biologically determined sex differences that incline but do not compel the individual's psychological development in the general directions of masculinity or femininity.

Sexuality is not synonymous with sex but is a construct comprising two components. One of these is gender role and the other is gender identity. Separable gender roles for males and females are defined and perpetuated by every human society and form the basis for a society's implicit definition of masculine and feminine. Gender identity is one dimension of the self-concept; specifically, it refers to the individual's concept of his or her own masculinity or femininity.

Gender identity develops gradually by successive stages that bear formal resemblances to those involved in the ontogeny of maleness and femaleness. Ontogenesis of masculinity and femininity depends upon learning and rests heavily upon language. The child must learn the essentials of masculine and feminine gender roles as defined by his society, and he must also learn to govern his own behavior in conformance with his assigned gender role. The degree of agreement between the understood gender role and the individual's perception of his own behavior and motivations determines the strength and constancy of his gender identity.

Ninety-nine percent of the time covered by the evolution of *Homo sapiens* preceded the domestication of plants and the invention of agriculture. In other words, for all but the last one percent of his existence man has been a semi-nomadic, group-living, hunter-and-gatherer; and these features of his history strongly influenced the evolutionary development of human sexuality. An increasing reliance upon animal food involved division of labor along sexual lines since the male's superior size, strength, and endurance equipped him for hunting. Hunting, killing, and butchering large animals necessitated planning, cooperation, and leadership.

The economic role of women in preagricultural society included gathering plant foods. Hunting males and gathering females brought their spoils to the group living place where food sharing was an absolute necessity.

Specialization of sex roles was also an adaptive necessity in the area of reproduction. Men had to impregnate women, and women had to bear and nourish children. Evolution of some type of nuclear family almost certainly occurred before or soon after the emergence of *Homo sapiens*. The values were both reproductive and economic.

One special factor that may have influenced evolution of the family was the disappearance of sharply delimited periods of behavioral estrus in the human female. One result was the extension of copulatory activity far beyond the female's fertile period and consequent necessity for an overall increase in the total frequency of heterosexual intercourse. It seems possible that family structure contributed to a heightened level of sexual interaction and that the increase in sexual relationships exerted a feedback effect strengthening selective heterosexual pair bonding.

The general conclusion suggested but by no means established is that as human social life became more and more complex, increasingly dichotomous specialization of social roles for males and females became more and more adaptive in maintaining and perpetuating the species and the human way of life.

The evolutionary changes envisaged in this interpretation occurred very gradually and over periods of hundreds of thousands of years. Their development was governed in part by the forces of natural selection, and their survival in the behavioral repertoire of the species depended upon their adaptive value to the species rather than to the individual. The evolution of separate gender roles also depended, of necessity, upon intergeneration transmission of culturally defined traits, values, and ways of behaving. Because behavior has feedback effects upon the gene pool of the reproducing population, sex differences in suitability for a specific gender role were favored by selection.

To state the argument in a single sentence, sexuality is a product of human evolution, and the evolution of *Homo sapiens* demanded as one essential component the progressive development of sexuality.

References Cited

August, C. P., M. M. Grumbach, and S. L. Kaplan
1972   Hormonal changes in puberty: III. Correlation of plasma testosterone, LH, FSH, testicular size, and bone age with male pubertal development. J. Clin. Endocrin. and Metab. 34: 319–326.

Beach, F. A.
1971   Hormonal factors controlling the differentiation, development and display of copulatory behavior in the ramstergig and related species. *In:* Biopsychology of Development. L. Aronson and E. Tobach, eds. Pp. 249–296. New York: Academic Press.

Chance, M., and A. Mead
1953   Social behavior and primate evolution. Symposia of the Society for Experimental Biology 7:395–439.

Crandall, V. J., and A. Rabson
1960   Children's repetition choices in an intellectual achievement situation following success and failure. J. Genet. Psychol. 97:161–168.

Davenport, W.
1965   Sexual patterns and their regulation in a society of the Southwest Pacific. *In* Sex and Behavior. F. A. Beach, ed. New York: John Wiley and Sons.

DeVore, I.
1965   Male dominance and mating behavior in baboons. *In* Sex and Behavior. F. A. Beach, ed. New York. John Wiley and Sons.

Eibl-Eibesfeldt, I.
1970   Ethology. The Biology of Behavior. New York: Holt, Rinehart and Winston.

Field, P. M., and G. Raisman
1973   Structural and functional investigations of a sexually dimorphic part of the rat preoptic area. *In* Recent Studies of Hypothalamic Function. K. L. Lederis and K. E. Cooper, eds. Basel, New York: S. Karger.

Ford, C. S., and F. A. Beach
1952   Patterns of Sexual Behavior. New York: Harper and Row and Paul B. Hoeber, Inc.

Goldschmidt, W.
1959   Man's Way. New York: Holt, Rinehart and Winston.

Gorski, R. A.
1971   Gonadal hormones and the perinatal development of neuroendocrine function. *In* Frontiers in Neuroendocrinology. L. Martini and W. F. Ganong, eds. Pp. 237–290. London: Oxford University Press.

Hammond, W. A.
1895   Cerebral hyperaemia: the result of mental strain or emotional disturbance. The so-called nervous prostration of neurasthenia. N.Y.: Brentano's.

Kagan, J.
1970   Personality Development. *In* Personality Dynamics. I. E. Janis, ed. New York: Harcourt, Brace and World.

Katcher, A.
  1955   The discrimination of sex differences by young children. J. Genet. Psychol.
         87:131–143.

Kephart, W. M.
  1973   Evaluation of romantic love. Medical Aspects of Human Sexuality 7:92–
         108.

Kohlberg, L. A.
  1966   A cognitive-developmental analysis of children's sex-role concepts and
         attitudes. *In* The Development of Sex Differences. E. E. Maccoby, ed. Pp.
         82–173. Stanford, California: Stanford University Press.

Kummer, H.
  1971   Primate Societies. Chicago: Aldine Atherton.

Masters, W. H., and V. E. Johnson
  1966   Human Sexual Response. Boston: Little, Brown and Co.

Mussen, P. H., J. J. Conger, and J. Kagan
  1969   Child Development and Personality. New York: Harper and Row.

Ramsey, G. V.
  1943   The sexual development of boys. Amer. J. Psychol. 56:217–234.

Sahlins, M.
  1959   The social life of monkeys, apes and primitive man. *In* The Evolution
         of Man's Capacity for Culture. J. Spuhler, ed. Detroit: Wayne State
         University Press.

Smith, M. E.
  1933   The influence of age, sex and situation on the frequency and form and
         function of questions asked by preschool children. Child Development
         2:201.

Stone, C. P., and R. G. Barker
  1939   The attitudes and interests of premenarchial and postmenarchial girls. J.
         Genet. Psychol. 54:27–71.

Tanner, J. M.
  1962   Growth at Adolescence. Oxford: Blackwell Scientific Publications.

Washburn, S. L., and C. S. Lancaster
  1968   The evolution of hunting. *In* Perspectives on Human Evolution. S. L.
         Washburn and P. C. Jay, eds. New York: Holt, Rinehart and Winston.

Zuckerman, S.
  1932   The Social Life of Monkeys and Apes. New York: Harcourt, Brace and World.

Beatrix A. Hamburg

The Biosocial Bases
of Sex Difference

Hamburg gives an overview of the biosocial bases for sex differences. Because of the complexity of interplay between biological endowment and environment, a very wide variety of lines of evidence have been sampled: evolution, animal behavior, genetics, endocrinology, embryology, and developmental studies. The brain matures in an environment that is different for males and females —social rewards and costs are not the same for the two sexes. Whatever insights may be gained from the past, it must be remembered that the modern world is really new.

INTRODUCTION

This chapter examines the origin and development of sexual dimorphism in human behavior. An evolutionary perspective is employed to provide a framework for integrating data from diverse fields of biology and the behavioral sciences.

A description of the modern synthetic theory of evolution is given with emphasis on behavior as adaptation. Probable selection pressures for the enhancement of sex differences in the behavior of early hominids are described. Categories of behavior believed to have been critical for survival in ancient times are examined for sex difference in modern men and women. These categories include: reproductive behavior, aggression, language, spatial behaviors, and affectional systems (pair-bonding) between adults and between adult and child. Concepts of attachment and dependency are discussed in connection with affectional systems.

Consideration is given to developmental, experimental, and comparative studies that have been drawn from child development, experimental psychology, sociology, anthropology, molecular genetics, neurology, primatology, ethology, neuroendocrinology, neuroanatomy, and the medical sciences. In this overview there will be an effort to describe the relevant research rather than to attempt to be exhaustive. For those desiring greater detail, references are given for recent comprehensive reviews in each area.

Throughout, terminology referring to masculinity and femininity is used. It is important to recognize that these are modal terms that refer to behaviors that are performed characteristically or more frequently by one sex rather than the other. The only behaviors performed exclusively by females are those that are linked very directly with childbearing. Ejaculatory behavior with discharge of spermatozoa is the only exclusively male behavior. It is important to emphasize this overlap in masculine and feminine attributes because there can be a tendency to think in terms of polarized sex stereotypes. Unfortunately, in the past, even some of the scientific discussion about roles of men and women has been clouded by partisan emotionality and, perhaps, an important misconception. Dobzhansky (1969) has addressed this latter issue very well. He was referring to racial differences but the principle is equally valid for sex differences. He said, "much honest, and some not so honest, confusion arose because of the failure to distinguish equality from identity and diversity from inequality. Equality is a political, ethical and religious precept; diversity is a biological fact . . . Diversity is compatible with equality of opportunity" (p. 45). It is now a prevailing view that equality of opportunity should be an inalienable right of all individuals regardless of what science may or may not reveal about sex difference.

It is generally acknowledged that despite varying and at times contradictory specifics of behavior, one of the universals across cultures has been the fostering of distinctiveness between the sexes. It may not be coincidental that deviations that tend to diminish the sexual dimorphism are less acceptable than those that enhance the differences. For example, short, weak men and large, powerful women are both stigmatized. Idealized males and females are the most stereotypical. In general, women have functioned in domestic roles as mother and homemaker and men in more public and prestigious roles in occupations, politics, and as warriors (D'Andrade 1966; Rosaldo 1974). Social change related to industrialization has given impetus for some change in traditional sex-stereotyped roles in the United States and concomitant changes in the social institutions that in the past have acted as constraints to maintain women in traditional roles.

There have been advances in biomedical technology that have improved the health and longevity of women relative to men. There are major advances in the technology of contraception. At the same time economic development has increased the demand for women in the labor force (Oppenheimer 1973). Significant changes in values are reflected in public policy. There are liberalized laws about abortion, public support for widespread, low-cost day care centers, and legal mandates for acceptance of women into a broader range of work and career roles. New patterns of behavior are emerging for women of all social classes. (Women of the poorer classes have always had to work outside the home in large numbers.)

U.S. Census data (1971) reveals a current trend toward later age of first marriage and an unexpected sharp drop in birth rate. Over the past two decades there has been greater acceptance of married women in work roles outside the home. In 1948, 13 percent of mothers with children under six years of age were in the labor force. By 1969, this figure had nearly tripled and 30 percent of such mothers were working. In 1948, 31 percent of mothers with children over six years of age were working. By 1969, 51 percent of such mothers were employed (White House Conference on Youth 1970). Older women are also working in greater numbers. In 1970, between 49 percent and 54 percent of women in the 35–59 age groups were in the labor force (Rosaldo 1974, Ginzberg 1971). The net effect has been an accelerated family cycle in which parenthood is a relatively short interval in the early phase of a greatly lengthened life span and there is a tendency for women to be employed for a large portion of their adult life. Recently among young people there has been interest in a wide range of communal family types and in revisions of family arrangements to de-emphasize the sex-typed division of roles in household labor. They challenge the hierarchical, patriarchal family roles that have been traditional.

Despite the current tendency to focus on women, there is also an important need to scrutinize the roles of men. There are social indicators of role strain in males (Komarovsky 1973) and evidence of higher casualty rates for males in medical and psychiatric disorders. It is valuable to examine the fit between biological predispositions in the male and the contemporary sex-typed roles and cultural expectations for men.

Basic Terms

Sex Role. Individuals are dichotomously assigned as male or female at birth based on observed characteristics of the external genitalia. When this occurs, a sex role is also being ascribed. Keller (1972) describes sex role as a master role that serves as organizer of the prescriptions and prohibitions for all spheres of functioning. Not only are attitudes and behaviors of individuals defined by the sex role, but the reciprocal behaviors of others towards them, are also defined.

Core Gender Identity. The phrase core gender identity refers to the recognition, as part of one's self-concept, of status as male or female. Money (1968) has found that this can be firmly established between 18 months and 3 years of age as judged by problems in sex reassignment for medical reasons. Biller (1971) uses *sex role orientation* to refer to the same phenomenon. In their usage, both emphasize the person's private view of the role. Biller makes other useful distinctions. *Sex role preference* refers to the individual's relative desire to adhere to culturally defined sex role guidelines and his evaluation of sex-related activities and interests. Usually sex role preference is consistent with sex role orientation (core gender identity). A split may occur in children if the core gender identity is affirmed by parents but they do not value traditional sex-typed activities. *Sex role adoption* refers to the overt masculinity or femininity of the individual's publicly observable behavior.

Biosocial. The term *biosocial* was selected in order to emphasize the interrelatedness of the biological substrates and the social environment as determinants of human behavior. There is no intention to ignore the influence of physical environment. In general, however, its role is already appreciated. It is widely recognized that the gene pool of an individual sets a range for defining the potential for the expression of a given characteristic, the *genotype.* Environmental factors weigh very heavily in the actual form attained, the *phenotype.* In the example of height, the phenotype or attained height will greatly depend on such environmental factors as health and diet. It should be noted that in the phenotypic expression of behavior, even those with a clear biological base, social and cultural factors also play a vital role. Indeed, even in the example of height, the motivational factors and interpersonal transactions of the individual could have a significant effect on health behaviors and dietary practices and thereby also influence bodily growth and ultimate stature.

It is precisely because of the complexity of interplay of endowment and environment that data from diverse disciplines in the traditional biological and behavioral sciences must be integrated to begin to truly comprehend human behavior. The important issue is not what biological behavior is exhibited per se, but rather, what is the biological contribution to the shaping of what is learned? We are interested in what may be the sex differences in the aptitudes, attentional preferences, temperament or motivations of males and females. Do males and females differ in the ease of learning of certain kinds of behavior? (Hamburg 1963). Are there factors that are systematically biasing the phenotypic behavior? A small systematic bias over time could ultimately have a large effect. If such biases do exist, what are the socioenvironmental factors that suppress or enhance their expression?

In assessing the biological components of human behavior, differentiation must be made between the individual and a population. For populations, most biological attributes occur in a normal distribution curve. Men and women are known to have distinctive but overlapping distribution curves for biologically determined physical characteristics, such as height and strength. Adult men are taller and stronger than adult women. The fact that a particular male may be smaller or weaker than most women does not invalidate the generalization. Biologically based sex differences in behavior will also occur in overlapping distribution curves. The behavior of a given individual cannot be evidence to prove or disprove a generalization for the population.

Although there is general recognition of the importance of learning in humans, there is, at times, a tendency to perceive biological behaviors as being unlearned, independent of the environment, and not readily susceptible to change. Associated with this there may be an expectation that there will be uniformity of response to a specific stimulus. For human beings, prior experience, motivation, and context will influence responses. It is hoped that the use of the term *biosocial* will serve as a reminder of the complexity and interactive nature of the processes involved in the development and expression of human behavior.

THE SYNTHETIC THEORY OF EVOLUTION

Simpson (1958) has given an excellent discussion of the theory that modern evolutionists have derived as a new synthesis from all fields of biology. It is a powerful instrument for organizing our thinking about biosocial issues. There are several basic concepts. The theory states that there is a heritable genetic code that sets reaction ranges within which the organism can develop. The characteristics of any individual organism are determined by the interaction of its heredity with its environment. Changes in this genetic code are continuously and randomly occurring as a result of the processes of mutation and recombination of the genes. Natural selection operates on these random changes to preserve those mutations and combinations in the gene pool that have been favorable. Favorable, in this context, refers to success in reproduction, or the enhancement of the likelihood that increased numbers of progeny will selectively survive and lead to continuation of those mutations in the gene pool of the population. This theory accounts for both novelty and stability in organisms since it deals with genetic changes that occur very slowly, over millions of years, in some cases.

159

The emphasis on behavior as adaptation is a central element. It not only underscores the importance of genetic factors in the explanation of existing behavior patterns, but more importantly, it involves behavior as one of the major factors that shapes evolution. Phases of selection that relate to breeding, care of the young, and coping strategies for survival are preeminently behavioral and are crucial elements in selection.

In taking this evolutionary perspective, it is also important to recognize that in large part our gene pool derives from a very ancient heritage. The primate order has existed for over 65 million years; primitive hominids have existed and have been evolving for two to three million years. Our own species, *Homo sapiens,* has existed for about 40,000 years. Washburn (1973) estimates that *Homo sapiens* have spent more than 99.5 percent of this span of time in hunting-and-gathering societies that were remarkably successful. Our most recent genes undoubtedly derive from this era. The advent of agriculture was about 10,000 years ago. It is believed that its world-wide adoption, in effective form, required several thousand years. Although the full impact of the technological advances on a worldwide basis has not yet been fully attained, the western world has had an industrial society for only about two hundred years. The postindustrial society is largely in its formative stages and even in the most advanced countries, it was initiated within the past 20 years. It is unlikely that there has been sufficient time to incorporate genetic changes from the agricultural or industrial eras in a widespread, stable way into the human gene pool. However, adaptations from the hunting-and-gathering era, and earlier, have given modern man a remarkable ability to systematically effect significant changes in his environment chiefly through tool technology, language, and refinements of social organization. Still, it must be borne in mind that westernized human beings are living in a modern technological world but equipped with an ancient vertebrate-mammalian-primate heritage deriving mainly from the adaptations that were appropriate to much earlier times.

EARLY MAN—NATURAL SELECTION PRESSURES FOR SEX DIFFERENCE

The importance of trying to understand early man has been well stated by Washburn and Lancaster (1968):

> human hunting . . . is a way of life, and the success of this adaptation (in its total social, technical, and psychological dimensions) has dominated the course of human evolution for hundreds of thousands of years. In a very real sense our intellect, interests, emotions and basic social life—all are evolutionary products of the success of the hunting adaptation. When anthropologists speak of

the unity of mankind, they are stating that selection pressures of the hunting and gathering way of life were so similar and the results so successful that populations of *Homo sapiens* are still fundamentally the same everywhere. [p. 213].

A relatively new and very active field of paleoanthropology has emerged and has led to advances in our understanding of the biology and culture of the early human being. It has drawn on the disciplines of vertebrate paleontology, evolutionary biology, and anthropology. Ethology has also made valuable contributions in terms of information and conceptualizations that relate environmental stimuli and phylogenetic adaptation to predictable behavior patterns within and across species.

It is significant that, as man, the hunter-gatherer has evolved, there have been selection pressures for the physiological and behavioral adaptations that enhance sexual dimorphism and sexual division of labor. Big game hunting was largely a male occupation (Steward 1936; Berndt and Berndt 1964, Spencer and Gillen 1927). Small game hunting, food gathering, and child care were done by females. These female occupations, it should be noted, contributed in a major way to the total food supply, frequently more than 50 percent.

There are sexual dimorphisms in physiological function, adaptations that confer male advantage in strength and endurance. For example: the grip strength of adult males is almost twice that of adult females (Damon et al. 1966); males have a proportionately greater vital capacity than females (Ferris and Smith 1953); males have a slower resting heart rate and greater stroke volume than females (Iliff and Lee 1952); males have a higher capacity for neutralizing the metabolite of exercise, lactic acid (Tanner 1970). Testosterone has anabolic effects that facilitate muscle and bone development. It enhances the synthesis of protein from amino acids and the retention of nitrogen, potassium, phosphorous, and calcium (Andersen 1966). The evidence for behavioral effects of testosterone in increasing activity level and aggressiveness will be discussed. These testosterone effects provide motivational and emotional complementarity to the physiological adaptations for a hunting life. All of the above-mentioned effects are notably exaggerated following puberty.

In going from life in the trees to ground-dwelling hunting life, much of the heritage from arboreal primate ancestry was highly useful. The *grasping* adaptations of the hand for climbing were useful in tool use and toolmaking. The binocular vision and excellent spatial intelligence were both readily transferrable and highly adaptive for the hunters. The new, hominid advances were in brain size and function. The most notable advance was undoubtedly the acquisition of language. Communal hunting would have been made much more effective by a highly developed communication system to facilitate planning and cooperation during the hunt and discussion afterward.

There was also a premium on slow maturation of the young and learning over a period of many years. Both hunting and gathering involved nomadic migrations over long distances, seasonally. Efficiency in exploiting each terrain would depend on detailed knowledge of the flora and fauna of diverse areas, knowledge of annual cycles and recall of unusual circumstances or rare events. None of this would be possible without highly developed intelligence and a long-learning, dependent childhood. There was a powerful selective advantage for maternal and child behaviors that fostered strong and prolonged attachment.

The evolution of the dominance hierarchy in primates has permitted the coexistence of multiple sexually mature males within the same territorially autonomous social group. In the evolution of the genus *Homo sapiens,* the exclusive and stable consort relationships between a male and one or more socially-bonded females has further diminished sexual competition among males. Some believe that it has contributed to the development of a level of cooperation between males which is unique.

There are also enduring mammalian-primate adaptations which foster male-male cooperation. Hierarchical status has been a major vehicle of social interaction and stability. In a social hierarchy system, there is usually little overt attack or injurious aggressiveness within a nonhuman primate group except under unusual stress or times of disequilibrium when new dominance ranks are being established.

Selection pressures have been exerted on the physical and behavioral attributes that are important in signalling status, threat, and appeasement. Selection pressure has favored sexual dimorphism in these attributes. For the female, this dimorphism would appear to have served to protect her from the likelihood of agonistic encounters. Guthrie (1970) makes a persuasive argument that virtually all of the secondary sex characteristics in humans serve a social signal function and that the sexual dimorphisms in physical appearance have been selectively enhanced. Females show an absence of structures and a minimizing of behaviors that serve as aggression elicitors or threat signals in the male. This is called complementarity of signal. Children strongly share in this complementarity of signal. In fact, many of the adult female characteristics at issue are pedomorphic. That is to say, they tend to resemble the childlike, nonthreatening form. There are many examples. It can be seen in the child and female absence of heavy chin and beard. In societies who do not shave, faces are a very visible and quick clue to age and sex status. In Guthrie's view and that of Livingstone (1967), residual hairiness of chest and shoulders in human males is related to the primate pattern of piloerection of these hairy areas in threat display. The relative absence of body hair in human females is interpreted as a complementary signal. The same analysis is given for differences in the quality of the voice, facial and jaw structures, musculature of the neck, breadth of the shoulders, and scalp hair patterns in adult males and females. Guthrie also discusses the positive relationship between threat display and courtship behaviors. He believes that there has probably been an inherent selection pressure for this association. Dominant males have more control over resources. Females who are attracted to and associate with dominant males could be expected to have more and healthier offspring.

In nonhuman primates hierarchical social structure has proven useful in monitoring the use of available resources. However, in nonhuman primates there is virtually no cooperation among group members in the collection of food and very little sharing of food among mature members. In feral chimpanzees the rare examples of food sharing are nonrandom and occur along kinship lines (McGrew in press). Feral chimpanzees have a rudimentary preadaptation in which males predominate in hunting, killing, and eating of mammalian prey. Females predominate in fishing for termites and driver ants (McGrew in press).

The social organization of human hunting-and-gathering society stands in marked contrast to the social adaptations of most other primates. It depends on group cooperation and food-sharing, both male-male and male-female, in a coordinated way that goes far beyond any other primate species. It also depends on the existence of an agreed upon home base to which both males and females return with their kills or food collections (DeVore and Washburn 1963).

The genetic predisposition for the development of intense and enduring affectional bonds between a mother and her offspring are part of the nonhuman primate heritage. It seems likely that in the evolution of early hominids some of the behaviors that underly the development of enduring adult male-female bonding were derived from adult-child nurturing behaviors. In the hamadryas baboon Kummer (1968) describes a sequence of development of an adult male-female bond arising from a transferred mother-infant relationship. Maternal, nurturing behaviors of feeding and grooming are suitable for enhancing social contact between adults. Stroking, patting, kissing, and stimulation of the breast are other examples of behaviors that occur in both mother-child and intimate adult-adult relations (Newton 1973).

The ability of the human female to experience orgasm comparable to the male enhances the reward value for both. It maximizes the utility of sexual behavior as a potent form of interpersonal bonding. In most nonhuman primates, adult male-female consort pairs are usually temporary and limited to a period of sexual receptiveness at estrus (gibbons are an exception). Stable adult human male-female pair bonds also may be facilitated by the absence of estrus and potential for sexual arousal and copulation at all stages of the menstrual cycle (Beach 1975). In any case, tradition in a particular society dictates whether there will be one or many wives, but an enduring relationship between the partners is the general rule.

It is important to restate that the characteristics and behaviors under discussion were those that had salience and were adaptive at the time of early hominids. It may be enlightening to look across contemporary cultures and see how universal or consistent are sex-typed roles and division of labor for comparison with ancient men and women.

Cross-Cultural Studies of Sex-Role Typing

Masculinity and femininity are institutionalized as statuses in all cultures. In addition to gender being used as a basis for assigning occupational tasks, it also serves as a basis for organizing social institutions.

A comprehensive review and analysis of cross-cultural data on sex-typed role behavior was done by D'Andrade (1966). He reported on data from many anthropologists and covered over 600 societies in terms of male-female division of labor, ascription of social status, patterns of interpersonal behavior, and phenomenology of gender identity. On the basis of the available data, he concluded that, although the specific behavior patterns are not universal, and, at rare times, are even reversed, there are modal typings of sex role and behaviors that are strikingly widespread. The prevalent finding is that males are more sexually active, more dominant, more deferred to, less responsible, less nurturant, and less emotionally expressive than females. Women almost universally were given child-rearing roles. Division of labor by sexes was almost universal. In general, male occupations tended to involve behavior that was strenuous, cooperative, and tended to require long periods of travel or absence. Tool making and weapon making, although not involving more strength or skill than manufacture or repair of clothing, were also male occupations. Tool making and weapon making appear to be activities that are assumed by men because of their use in relationship to activities defined as masculine. Women have major responsibility for gathering fuel, water, and foods. At times these gathering activities can be very strenuous and involve heavy burdens. Women also manufacture and repair clothing. In agricultural societies, men and women often perform equal work in the fields. Warrior roles are exclusively assigned to adult males.

In general, cultures are organized around males rather than females. In fact, D'Andrade points out that the institutional subordination of women is more pervasive and complete than can be explained solely by overt differences in dominance or aggression. He notes the sex bias that leads to a devaluation of female activities. For example, activities performed by women are evaluated less highly simply because they are performed by women.

Rosaldo and Lamphere (1974) have edited an anthropological volume in which 16 authors explore possible explanations for the universal subordination of women to men. In one way or another, the authors focus on issues of power and authority as related to participation in the domestic and public domains of a society. The biological linkage of females to childbearing and child-rearing are seen as having bound women to the domestic sphere. Sanday (1974), using data from the *Ethnographic Atlas,* gives an analysis of the proportionate expenditure of males and females in the survival activities of reproduction, subsistence, and defense (warfare). Although she hypothesizes that increased female contribution to subsistence should increase female status, the data does not bear this out. In over 50 percent of the societies cited, women contributed 50 percent or more of the subsistence and were still subordinate. LeVine (1970) has reported a similar finding when he studied effects of extended male absence on female subsistence contributions. The female contribution increased markedly but there was no elevation of status. It was mentioned previously that in hunting-and-gathering societies, women are calculated to have contributed at least half of the subsistence.

The role of male dominance and aggression in subordination of females is not discussed in Rosaldo and Lamphere (1974) and is minimized in D'Andrade (1966). The dimorphism in size is relatively small. However, the dimorphism in strength is great and has been underestimated (Damon et al. 1966). Furthermore, the disparity between the sexes may be greatly exaggerated when behavioral factors such as higher male aggressiveness and greater female vulnerability when she is holding or nurturing a child are considered. Female intimidation by adult males has been a part of primate adaptation. There is reason to believe that one should not underestimate the evolutionary and historical role of wife intimidation in human societies. It is extremely likely that coercion will become the exception rather than the rule in modern societies. There are increasing cultural as well as legal constraints on the use of force with both women and children. The subtle intimidation of economic dependence is being negated by increasing female employment and the economic supports by the welfare state.

To appraise current sex-role stereotypes in the United States, Broverman et al. (1972) conducted a study that cut across lines of age, sex, marital status, and education. They concluded that there is strong consensus across all groups as to the sex roles of men and women. The characteristics ascribed to men reflect a *competency* cluster. Men are described as independent, objective, active, competitive, logical, skilled in business, adventurous, decisive, self-confident, ambitious, and taking leadership roles. The stereotypic perception of women is as dependent, noncompetitive, passive, gentle, warm and expressive, sensitive to feelings of others, and able to express tender feelings. Characteristics ascribed to males are more positively valued than the traits ascribed to women. In general, the sex-role definitions are implicitly and uncritically accepted and incorporated into the self-concepts of both men and women. These stereotypes are considered desirable by all groups, even college students who are often critical of traditional social norms. This research confirms the widespread acceptance in contemporary United States of the same sex role behaviors and typings that have been postulated for early hominid societies and those that appear to have characterized most of the societies studied in cross-cultural surveys.

REPRODUCTIVE BEHAVIOR

General Considerations

The chromosomal, hormonal, neuroendocrine, and behavioral influences on reproductive behavior will be considered. There are several detailed, analytic reviews of these topics (Bardwick 1971; Hutt 1972; Money and Ehrhardt 1972; Friedman et al. 1974).

Reproductive success of the population ultimately depends on the kinds, frequencies, and combinations of genes and chromosomes that are in the genetic pool. Sexual, as compared with asexual, reproduction increases the diversity of a gene pool. The selective advantages of diversity in the gene pool have been amply demonstrated. On the one hand, heterozygosity offers the possibility of offsetting the potential negative effect of a particular mutant through the presence of a normal allele. There is also the advantage of diversity in enabling a population to show greater flexibility in responding to environmental change or opportunity. When an environmental change becomes stable and a particular genetic variant is notably advantageous, the trait will be passed on selectively and, over time, become incorporated as a stable element of the genetic makeup of the population. In times of rapid change, a population with a large and diverse gene pool has *preadapted* individuals who are genetically prepared for a range of circumstances. Structural and behavioral mutations that have facilitated sexual reproduction have, therefore, had an adaptive advantage and have been favored in the natural selection process.

Sexual, biparental reproduction is fundamentally dimorphic at all levels. Each parent contributes an homologous but distinctive set of chromosomes to the zygote (fertilized egg). Anatomy of reproductive organs and patterns of sexual behavior must be sufficiently dimorphic to permit fertilization to occur. Adults need to be sufficiently dimorphic in nongenital structures and behaviors that they can be identified and be located as members of the opposite sex with relative ease.

Chromosomes and Sex Difference

Much is known about the mediation and sequence of events culminating in differentiation of male and female gonadal structures. The XX (female) or XY (male) chromosomal configuration functions chiefly to determine the differentiation of the embryonic genital anlagen into ovary or testis. The Y-chromosome is prepotent in this function and the effect of the Y-chromosome is present even in such combinations as XXY, or even XXXXY (Gorlin 1965). Conversely, in the absence of the Y-chromosome the individual always differentiates as a female, even in the instance of a single X-chromosome such as the XO configuration of Turner's syndrome.

The Y-chromosome is believed to be responsible for the elaboration of an *organizer* substance which causes the inner (medullary) portion of the primordial genital ridge to form a testis. In humans this occurs very early in embryonic life (seven weeks). Testosterone also has other important influences on the developing brain that will be discussed under hormonal determinants of behavior (page 170).

The relatively recent advances in cytogenic techniques for studying mitotic chromosomes has yielded a great deal of information about sex-chromosome function and anomalies. The function of the second X-chromosome has been partially elucidated. It is now known that the second X is required for full oogenesis. In the XO condition the ovary initially develops normally in fetal life but starting at three months of gestation there is aplasia and, at birth, there is a striking loss of germinal cells. By puberty, no ova are detectable. The second X seems to have other more general effects. The XO individuals are of unusually short stature. Exclusively feminine attributes have been described for all of these women, (Ehrhardt, Greenberg, Money 1970) based on measures of tomboyism, energy expenditure levels, and preferred clothing and interests. Also, a profound deficit of spatial perception has been recorded (Alexander et al. 1966) in a large proportion of cases.

On the male side, there is continued extensive study of 47 XYY males. These males have been associated with unusual tallness. The relationship of XYY to aggressive, criminal behavior is under scrutiny although it now seems unlikely that there is a positive correlation as believed earlier (Kessler and Moos 1972, 1973, Meyer-Bahlburg 1974).

In the Klinefelter syndrome males XXY, male development is incomplete. Testes fail to enlarge normally at puberty. These individuals seem to be unusually prone to mental deficiency and emotional instability (Money and Pollitt 1964; Ferguson-Smith 1966).

X-Linked Conditions

There is a well-documented major difference in the morbidity and mortality of males and females that appears to be partially related to the chromosomal difference. At all ages males show greater vulnerability and have a higher mortality. This surplus of male death continues to exist despite advances in medical and surgical technology. This greater vulnerability is true for all mammals that have so far been studied (Wolstenholme and O'Connor 1962).

It is worth noting and probably significant that there is a quantitative difference in the actual amount of chromatin material in X and Y chromosomes. The Y chromosome is much smaller, a mere fragment of the X chromosome. The Y chromosome carries very few genes, if any, other than those that determine male sex. Many genes are carried on the X chromosome. The disparity in size and function of the X and Y chromosomes means that the homogametic, female sex is diploid (XX) while the heterogametic, male sex is haploid (XY). Childs (1965) has stated:

> The differences so created would be due to dosage effects compensated incompletely or not at all (in the males) or in the presence of compensation to the opportunities for heterozygosity in females which are denied to the male. Some of these differences are very great, as in the case of some conditions in which the presence of a normal allele in the female protects her against the ravages of a mutant which, unopposed by any homologue in the male does harm . . . No doubt some of the genes in the X chromosome are so placed because there is some selective advantage in their presence there, but some may have been left by the accidents of the evolution of sex determination. It is possible, therefore, that some of the biological superiority of the female may be a happy concomitant, and the burden of disadvantage which is the heritage of the male, an unfortunate by-product, of the evolution of a workable means for maintaining variability through sexual reproduction. [p. 810]

The linkage with sex offers opportunities for chromosome mapping. As research in this field has progressed, differentiations are now being made between sex-linkage and sex-limitation. *Sex-linkage* refers to those conditions that are carried on the X-chromosome. Known X-linked diseases are Vitamin D-resistant rickets, nephrogenic diabetes insipidus, a number of drug-induced anemias, immunoglobulin deficiency, clotting disorder, color blindness, and hypoparathyroid deficiency. In these conditions heterozygous females are protected by an isoallele and escape illness entirely or have a mild condition due to the ameliorating effect of the isoallele, whereas in males the disease expresses itself in full-blown form. Another characteristic of sex-linked conditions is that the frequency of occurrence is greater in males than in females.

Hormonal Determinants of Sex Difference

Initially the embryo is undifferentiated. Under normal conditions the chromosomal endowment (XY or XX) sets a program that entrains sequential events leading to progressive stages of differentiation of internal and external reproductive anatomy to definitive male or female structures. These are mediated through hormonal effects on the developing embryo at critical periods. There are also hormonal effects on the developing brain that lead to neural organization for the male or female patterns of cyclicity, sensitivity to gonadal hormone level, and patterns of sexual behaviors to be discussed on page 174). Experimental and clinical aberrations of the normal sequence of development will be discussed in terms of their effects on sexual anatomy and behavior.

Differentiation of the Gonads

In fetal development, initially there is an undifferentiated genital ridge in the embryo that can proceed in either a male or female course of development. The inner medullary portion has the potential of testis development. The outer (cortical) portion of the ridge has potential for ovary development.

Jost (1972) has noted that there is a sex difference in the timing of the development of the fetal gonads that is significant. Under the influence of the Y-chromosome the testes develop very early in the male. In human beings, the development and organization of the seminiferous cords in the testis occurs in the seventh week of gestation. The ovaries are much slower in differentiating. Jost hypothesizes that there is no induction effect on the cortical portion of the genital ridge in either sex until a timing later than that for development of the testis. In males, the newly formed embryonic testis produces male gonadal hormone, testosterone. It also produces another substance that acts locally to inhibit the differentiation of female structures such as the Mullerian ducts (Jost 1953; Federman 1967). In females, in the absence of testosterone, the gonadal primordium slowly becomes an ovary. The female hormone, estrogen, does not serve as a feminizing trigger. Experiments in animals have shown that equally good development of female structures such as uterus and fallopian tubes occurs in the absence of the fetal ovary as long as no testosterone is present (Jost 1958). Development of male structures depends on testosterone.

The external genitalia develop from the same primordia in both sexes, a genital tubercle and uro-genital slit. After the eighth week of fetal life, testosterone promotes the development of a penis and scrotum. In the absence of testosterone, a clitoris and labia majoria differentiate.

Sex Hormones and Neuroendocrine Function

The details differ somewhat between species but sexual differentiation of the brain is a basic mammalian pattern. Detailed recent reviews include Saunders (1968), Whalen (1968) and Gorski (1971).

Sex differences in the pattern of gonadotropin release from the pituitary was first reported by Pfeiffer (1936). It has been subsequently found that locus of regulation and the site of sex differentiation is the hypothalmus and is dependent on the presence or absence of testosterone at a critical time in fetal development.

The female is characterized by cyclic activity. Gonadotropic hormone peaks dramatically during the preovulatory period. This regular peak in hormone activity results in cyclic ovulation, a readily detected cycle in the vaginal epithelium and cyclic changes in arousability or sexual receptivity.

In contrast, the male is noncyclic in gonadotropin release. The experimental production of a female pattern in genetically male animals is best achieved by total removal of testicular androgens. The administration of estrogen, progesterone or other variant forms of gonadal hormones, leads to variable outcomes that can be paradoxical or indeterminate depending on the timing, dosage, and site of administration. In prenatal hormonal effects, Money (1972) states, the antithesis of androgen is not estrogen, but nothing.

The androgen effect on the hypothalamic system depends on timing. Androgen administered either before or later than the species-specific critical period is ineffective. During the critical period the brain is sensitive to very small amounts and brief exposure to androgen. For example, in rats, Arai and Gorski (1968) found that a six-hour exposure to androgen in females within the first week post-partum was sufficient to cause the acyclic, male pattern of gonadotropin release. In rats, the critical period for hypothalmic cyclicity peaks a day or two earlier than the critical period for hypothalamic regulation of dimorphism in sexual behaviors.

The pituitary gland secretes the gonadotropic hormones but itself does not become sexually differentiated. This was demonstrated by exchange transplantation of mature pituitary glands between males and females.

When the transplants were properly located in contiguity with the hypothalmus of the operated animal, prior release patterns were maintained. Females with transplanted male pituitaries continued to show a normal female cyclic pattern. Males with pituitary transplanted from a female continued to show a noncyclic pattern (Harris and Jacobsohn 1952; Martinez and Bittner 1956). It was further confirmed that transplant of pituitary could not reverse the experimentally induced noncyclicity of females whose functioning had been altered by injections of androgen at the critical period (Segal and Johnson 1959).

Hypothalamic neuroanatomy and neurophysiology have been extensively studied to elucidate the sites and mechanisms of the neural regulation of gonadotropin release. Evidence from a variety of techniques such as electrolytic destruction, hormone implantation, electric stimulation, and autoradiography has been reviewed (Lisk 1967, Green et al. 1969, Gorski 1971). Investigators believe that the arcuate (tuberal) region of the hypothalmus regulates the tonic (noncyclic) discharge of gonadotropin from the pituitary. The preoptic area is believed to regulate the cyclic release system. Halasz (1969) has some evidence that the trigger for spontaneous ovulation may reside in the preoptic area.

Sex differences in the ultra microscopic structure of the hypothalmus have been found (Raisman and Field 1971). The axons of the stria terminalis differ in their mode of termination in the preoptic and tuberal (arcuate) hypothalamus. Using electron microscopy, they showed that the number of nonamygdaloid dendritic spine synapses in the stria terminalis of the preoptic area is twice as high in the female as in the male. These investigators later demonstrated experimentally that the sexually dimorphic neuronal structures of the strial part of the preoptic area undergo neonatal differentiation under the influence of androgen in exactly the same way as the sexually dimorphic functions of gonadotropin release and sexual behavior (Raisman and Field 1973).

Sex Difference in Brain Regulation of Sexual Behaviors

This complex area has been well reviewed by Beach (1947), Whalen (1968), and Money and Ehrhardt (1972). Over the years, it has been established that under the influence of gonadal hormones at critical fetal or neonatal periods, the mammalian brain is programmed for sex difference in characteristic mature sexual behavior and copulatory patterns. Not only is the characteristic behavioral response determined, but the threshold point is set for later sensitivity of the brain to circulating gonadal hormone levels.

The basic strategy of research in this field is to study the later effects of a variety of hormonal manipulations of the fetal or neonatal animal. Animals are injected with gonadal hormones either as intact or castrated animals. Cross-sex hormone is often employed. Female sex hormones are injected either singly or in combination. Timing or priming strategies may be employed. At puberty the experimental animals may be untreated or injected with same or cross-sex hormones. Finally, the experimental animals may be behaviorally tested with the same or opposite sex partners, or both. Timing and hormone dosage effects must be controlled.

Environmental influence has been shown to exert organizing effects in utero that significantly alter adult sexual behavior. Using rats, Ward (1974) demonstrated that noxious stress to pregnant mothers "had little effect on female fetuses but it radically altered the course of sexual differentiation in the males. Males were behaviorally feminized and demasculinized" (pp. 10–11). In the same animals, there were no significant changes in genital structures or spermatogenesis. Ward postulates that the behavioral effects are due to a stress-related, partial prenatal androgen deficit. The differential response in morphology and behavior may be dependent on differences in thresholds of response to testosterone or the morphology and behavior responses may be mediated by different androgens.

Clemens (1972, 1974) showed an analagous threshold effect in females with a partial prenatal androgen excess. The androgen secreted by the embryonic testes of the male fetuses appears to reach the female fetus by diffusion across amniotic membranes. He found, in rats, that a female fetus developing in utero between male fetuses or in the presence of more than three male fetuses was likely to show signs of masculinization of mature sexual mounting behaviors even though the amount of prenatal androgen excess had not altered her anatomical structures.

It has been important in interpreting the data in sexological studies to pay careful attention to the differences within species. Beach has provided valuable paradigms for the analysis of dimorphic sexual behavior. He has developed detailed analyses of the components of characteristic male and female behaviors of the rat (Beach et al. 1969). This kind of conceptualization has been important in attempting to specify the many variables in behavioral studies and to quantify the results in reporting masculinity and femininity of response. Also, studies of mounting behavior have been confounded by the deficiencies in growth and sensitivity of the penis in neonatally castrated animals. These factors can independently affect the number of intromissions achieved (Beach 1968).

In general, the findings confirm the same pattern previously described for cyclicity of gonadotropin release. Differentiation of reproductive behaviors will proceed in a male pattern in the presence of testosterone at the critical period of fetal or neonatal life. It has been found that in instances in which a genetic male has been subjected to female hormone influence at the critical time, either masculine or feminine sexual behaviors can be elicited in adulthood depending on whether male or female hormones are administered. When genetic females are subjected to high doses of male hormone at the critical early period, the female component is usually rather completely inhibited.

Over the course of evolutionary development there has been increasing complexity of the neuro-endocrine-environmental mechanisms that control sexual behavior. Mature human sexual behavior represents a culmination of this trend and social learning appears to play a superordinate role. Comparative studies of guinea pigs (Valenstein and Young 1955), rats (Beach 1947), cats (Rosenblatt 1965), and monkeys (Luttge 1971) show a decreasing dependence on levels of sex hormones and an increasing dependence on cortical brain function with ascent of the mammalian scale. It should be noted that even in animals such as the rat, the sexual behavior displayed by experimental and control animals is significantly influenced by immediate prior experience. The cortical, experiential components of behavior cannot be totally ignored in any of the mammalian studies regardless of the ranking on the phyletic scale.

It seems reasonable to assume that the similarity to the neuro-endocrine control of gonadotropin release and the development of gonads and external genitalia is not accidental. There would be selective advantage in terms of high probability of fertile mating if the physiological, structural, and behavioral components of reproductive behavior were part of a coadapted genic complex.

SEX DIFFERENCE IN BRAIN REGULATION OF NONSEXUAL BEHAVIORS

Experimental Hermaphroditism in Rhesus Monkeys

Phoenix et al. (1968) studied fetal androgen influence on the sex differentiation of the brain in rhesus monkeys. This animal was chosen because of its close phyletic relationship to man and the knowledge that there was prominent sexual dimorphism in the play patterns of males and females (Rosenblum 1961; Harlow 1965). They wished to answer the question of whether or not prenatal gonadal hormones play a role in patterning the subsequent sexually dimorphic behaviors.

Pseudohermaphroditism was induced by introducing testosterone during the period of genital differentiation. Although the ovaries were intact and internal structures were female, these animals were all born with a small, normal appearing penis and a well-developed scrotum. The experimental animals displayed increased mounting behavior and male juvenile play patterns. They were high in threat, chasing, play initiation, and rough-and-tumble play. This behavior was observed and measured in normal males, experimentally induced hermaphroditic females, and normal females starting at three months of age and continued until the end of the second year of life. The pseudohermaphrodites displayed frequencies of performance of rough-and-tumble play that were usually intermediate between the high levels of normal males and low levels of normal females and sometimes even equal to the normal male standard. Control (normal) females did not at any time equal or exceed the average for the pseudo-hermaphrodites, much less attain the high levels seen in males.

HORMONAL INFLUENCE ON SEX DIFFERENTIATION IN HUMAN CLINICAL EXPERIENCE

For ethical reasons, it is not possible or desirable to conduct experiments with human beings that would lead to disturbances in sex differentiation. However, there have been some naturally occurring situations in which the hormonal influences in utero are drastically altered. These are, in effect, natural experiments that can be studied and that offer an opportunity to examine later behavioral as well as biological effects of high doses of testosterone in utero on genetic females and of comparable high fetal estrogens on genetic males.

These naturally occurring situations have arisen in several ways. The high fetal testosterone in genetic females has arisen in instances in which the individual inherits a condition characterized by defective functioning of the adrenal gland, the *adrenogenital syndrome.* Instead of producing the proper hormone, cortisol, the afflicted individual produces instead, a precursor product that is, in biological action, a male sex hormone. This androgen begins to enter the circulation too late in fetal life to induce extensive masculinization of the internal reproductive system (Wolffian ducts). However, it is in time to masculinize the external genital anlagen and, in the extreme form, lead to formation of a penis and fusion of the labia to form an empty scrotum, or, more commonly, to lead to an intermediate form characterized by an enlarged clitoris and partial fusion of the labia majora. In any case, the female infant is often mistaken for a male at birth.

A related syndrome of progestin-induced hermaphroditism was inadvertently induced in a small number of genetic female fetuses when hormones were given to the mothers to prevent miscarriage. These hormones, progestins, were of a recently synthesized group of steroids that had biological actions similar to the female hormone, progesterone. However, at first, it was not known that they could sometimes exert a masculinizing influence on a female fetus. When this was discovered the use of the hormone was discontinued, but in the 1950s there were some female babies born with significant hermaphroditism. The children with complete masculination of the external genitalia were raised as boys. The individuals with incomplete masculinization of external genitalia were more extensively studied and when it was learned that the genetic, gonadal and internal sex types were female, they were surgically repaired and designated as girls.

The hermaphroditic girls have been extensively studied from birth through childhood and adolescence (Money and Ehrhardt 1968, 1972) in order to determine whether or not prenatal hormones may have affected the fetal brain in such a way as to influence later behavior. In the progestin-induced cases, there was no further exposure to androgenic hormones after birth. The adrenogenital syndrome children were administered the appropriate hormone (cortisol-type) postnatally and in most instances, they, too, were without exposure to masculinizing hormones after birth. These girls and their parents were studied for such behaviors as tomboyism, energy expenditure, and aggressiveness, and also for clothing preferences, attitudes about childhood sexuality, maternalism, attitudes concerning careers and marriage, and romanticism. The fetally androgenized girls (experimental group) of both groups, when compared to matched controls, demonstrated greater tendencies to be tomboys, preferred more active, energetic activities, but they did not show an increased predisposition for overtly aggressive behavior such as fighting. The experimental girls were significantly low in maternal interests or interest in dolls and showed a striking lack of interest in marriage and child-rearing. They showed a clear preference for toys characteristically employed by boys, such as guns.

The researchers (Money and Ehrhardt 1971, 1972) postulate that the increased predisposition of prenatally androgenized girls to tomboyism is caused by a masculinizing effect on the fetal brain. They suggest that despite the fact that these individuals were assigned to and reared as members of the female sex, they still exhibit some masculinized behaviors as a result of prenatal androgen effects. These natural experiments in human beings are in accord, therefore, with the established animal experimental models.

This important work has been confirmed and extended in studies by Ehrhardt and Baker (1973, 1974). Using mothers and sisters as control, they replicated the preference for rough-and-tumble play, masculine playmates, and masculine toys as characteristic of prenatally androgenized girls. These preferences were perceived as secondary to a basic temperamental preference for rough-and-tumble play. These androgenized girls also showed low interest in dolls, maternalism, or marriage. A new feature of their work has been the inclusion of boys with a history of excess fetal androgen. When compared to their brothers, they showed exceptionally strong preference for vigorous, rough outdoor activities. They were not more prone than their brothers to initiate physical fighting and verbal aggression. In none of the androgenized patients was the behavior considered as abnormal or beyond the accepted ranges of individual difference.

A study of boys with prenatal exposure to high concentrations of female hormones was reported by Yalom et al. (1973). The experimental group consisted of 16-year-old subjects and 6-year-old subjects whose mothers had been given moderately high doses of estrogen and progesterone during pregnancy because the women were diabetic. These boys were compared with matched control groups: (1) sons of diabetic mothers who had no exogenous hormones during pregnancy, and (2) sons of nondiabetic mothers. In comparison with controls, the 16-year-old subjects gave self-reports of fewer aggressive acts, were rated as less assertive and lower in athletic ability on teacher ratings; the 6-year-olds differed from controls only in lower teacher ratings on assertiveness and athletic ability. The 16-year-old experimental subjects showed greater feminization effects than the 6-year-olds. Longitudinal study could reveal whether this effect is real or a sampling error; if real, whether it reflects more stringent standards for masculinity postpuberty, small incremental shifts towards femininity, or an abrupt change related to hormonal puberty changes.

MENSTRUAL CYCLE

The endocrinology of the human menstrual cycle has been exceedingly carefully studied for many years. There is probably more known about the cyclic events of the sexual hormones of the human female than any other species. Recent technical advances in hormone determination have made fine distinctions possible. Schwartz (1968) has described the complexities of the neuroendocrine controls and the characteristics of the pituitary-ovarian feedback system in an excellent review. For our purposes, however, a schematic outline of the ovarian hormone changes in relation to the phases of the menstrual cycle will suffice (Figure 1).

The average menstrual cycle is 28 days in length. The days are numbered starting with the first day of menstrual flow. The *menstrual phase* of flow lasts from four to seven days. Cessation of menstrual flow marks the initiation of the *follicular phase*. At this time the Graafian follicle develops in the ovary. *Ovulation* typically occurs in the 14th day (mid-cycle) when the Graafian follicle ruptures and the ovum is expelled. This is followed by the *luteal phase* which refers to the transformation of the follicle into the corpus luteum. The latter stage of the luteal phase is distinctive and is called the *premenstrual phase*. The premenstrual phase describes the three to five days immediately preceding the onset of menstruation.

Estrogen levels show a dual peak. Estrogen is produced by the Graafian

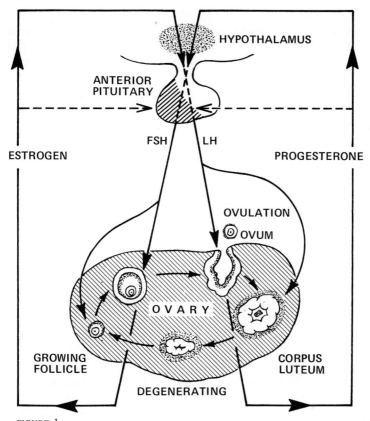

FIGURE 1
The menstrual cycle.

follicle and rises to a first peak at ovulation. It then shows a small dip but rises again quickly and there is a second peak at the mid-luteal phase followed by a rather sharp drop during the premenstrual phase.

Progesterone is produced by the corpus luteum. During the luteal phase, it rises from a very low level and peaks also at the mid-luteal phase. Like estrogen, progesterone drops sharply during the premenstrual phase. Both of the ovarian hormones, estrogen and progesterone, are very low during menstruation.

Many women do not experience cyclic mood or behavior change. There are, however, cyclic changes in moods and behavior that have been consistently reported in relation to phases of the menstrual cycle for some women. The pioneering studies in this field were done collaboratively by a psychoanalyst and gynecologist (Benedek and Rubenstein 1939) but were largely neglected until the past decade. Menstrual phase was judged by vaginal smear and rectal temperature. They found that during the follicular phase of the cycle, when estrogens are high, there are feelings of well-being with high physical and mental activity and extroverted heterosexual tendencies in behavior. At time of ovulation there was heightened sexual receptiveness and sense of sexual tension. Postovulation, the heterosexual tension is abruptly relieved and a period of relaxation follows. With the sharp decline in estrogen and progesterone levels in the premenstrual phase, Benedek and Rubenstein reported "regression in psychosexual integration" and described anger, fatigue, depression, low frustration tolerance, and generalized emotional lability. They also found that the onset of menstruation was usually accompanied by a relaxation of the emotional tension and irritability of the premenstrual phase, even though there could be an increase in physical symptoms and discomfort.

Although thorough replication of this work has not been completed as yet, such investigations as have been carried out tend to confirm these basic findings of mood cyclicity (Altmann, Knowles and Bull 1941; Gottschalk et al. 1962; Moos et al. 1969). Moos also called attention to the greater consistency of the activation, positive affects and sexual arousal variables of the intermenstrual phases as compared to the anxiety, aggression, and depression variables of the premenstrual and menstrual phases. He noted that high premenstrual tension women tend also to be higher in distress in other phases of the cycle as well.

The *premenstrual syndrome* was first recognized by Frank (1931) and he coined the term *premenstrual tension* to describe a triad of irritability, depression, and lethargy that are characteristically present. A wide variety of symptoms may be superimposed and the pattern of symptomatology for a given woman tends to be recurrent and stable although the intensity may fluctuate from cycle to cycle, often in relation to intercurrent environmental stress.

The premenstrual period has been associated with general somatic complaints; with edema, dermatitis, sleep disorder, vertigo, nausea, breast pain, and back pain (Dalton 1964).

Exacerbation of specific diseases during the premenstrual period have been reported for diseases such as asthma (Turiaf et al. 1949), glaucoma (Dalton 1967), migraine (Greene and Dalton 1953) and epilepsy (Logothetis et al. 1959).

Research studies have demonstrated specific premenstrual variations in EEG measures of arousal (Kopell et al. 1969), olfactory sensitivity (Vierling and Rock 1967), and capillary fragility (Clemetson et al. 1961) that seem to be related to some of the observed premenstrual symptomatology.

Reports of premenstrual effects on emotional reactions are even more striking. Numerous authors have reported correlation between premenstrual phase and (1) *suicide attempts* (Mandell and Mandell 1967, Glass et al. 1971, Wetzel et al. 1971, Ribeiro 1962); (2) *psychiatric emergencies*

(Jacobs and Charles 1970, Glass et al, 1971); (3) *admissions to psychiatric wards* (Dalton 1959, Kramp 1968, Mall 1958, Janowsky et al. 1969); (4) *crime* (Ribiero 1962, Dalton 1961), and (5) *serious accidents* (Dalton 1960).

Cross-Cultural Observations

Cross-cultural observations have been made of mood and behavioral changes of the menstrual cycle. Janiger et al. (1972) have made a cross-cultural questionnaire survey of six cultures (American, Japanese, Nigerian, Apache, Turkish and Greek). They found that the principal symptoms of the premenstrual syndrome are present in all the cultures studied although the frequency and severity of symptoms may differ. For example, the Japanese tended to report the fewest symptoms, Turks and Nigerians had the highest incidence, and Americans were intermediate. They plan further research taking into account cultural differences in menstrual taboos, women's social roles and attitudes, cultural norms with respect to pain response, and legitimacy of psychic and somatic complaints. They also sent questionnaires to primatologists and zookeepers asking about cyclic mood and behavioral changes in rhesus monkeys, chimpanzees, and gorillas. The communications from the Yerkes Primate Center and London Zoo revealed that the premenstrual phase in nonhuman primates observed by them was characterized by restlessness and irritability. These authors tentatively conclude that premenstrual distress is a universal phenomenon across human cultures and in relevant primate groups.

Several recent comprehensive reviews of the literature on studies of the behavioral correlates of the menstrual cycle (Sommer 1973; Parlee 1973; Rossi 1974) have raised methodological questions. There is concern about behavior indices that rely on retrospective reporting. This could lead to distortions that reflect the cultural stereotype rather than the specific individual experience. A publication bias was noted in which there was often failure to cite studies with negative results. Many studies rely on imprecise judgments as to the phase of the menstrual cycle. The underlying hormonal state was usually inferred instead of using actual hormonal assay. Very few studies were designed to measure hormone level and mood concurrently.

There is a well-designed study that avoids the errors cited above (Persky 1974). Persky found "remarkably little change in negative moods (anxiety, depression, hostility) across 29 cycles of healthy young women" (p. 464). The research design included radioimmunoassay of progesterone, testosterone and estradiol at early follicular, ovulatory, peak, and luteal phases. All subjects were given medical, psychiatric, and gynecological examinations. Psychological testing was done on each occasion of drawing blood.

Rossi (1974) has criticized the general assumption in menstrual studies that hormones are the cause of mood effects. Emotional factors have been shown to be the *cause* rather than the effect of gonadal hormone levels in some monkeys and man. These experiments are outlined by Rose et al. (1971:184).

Rossi (1974) has studied cyclicity of mood and sex behavior examining nonmenstrual factors. Data sets of daily self-ratings of single men and women students for mood (positive and negative) and sex behavior have been analyzed by phase of menstrual cycle, day of the week, state of the weather, interpersonal influence, and sexual activity. She found that for women there was firmer cyclic patterning of mood and sex activity by calendar week than menstrual month. Mood was most positive on weekends and least positive mid-week. In the menstrual cycle there was a weak tendency for a peak of positive mood at ovulation and a low point of negative mood during the first days of menstruation. Menstrual-related highs and lows were sharply influenced by synchrony with the day of the week (social unit of time). In other words, the highest moods were recorded when the ovulatory phase coincided with the weekend. There was some clustering of sex activity at the mid-menstrual cycle but half of all sex activity occurred on the weekend. Coital activity was a mood stabilizer for men but not for women. These data highlight the interaction between psychosocial and physiological variables in determining the mood and behavior of these individuals.

Oral Contraceptives and Cyclic Mood Change of Women

The oral contraceptives currently in use represent a wide variety of types of synthetic hormones, dosages, ratios, and sequencing of the pills. Hamburg and Lunde (1966) have reviewed the evidence for differential effects of the hormones depending not only on absolute levels of hormone but also on interactional effects such as prior priming of progesterone by injections of estrogen and, at times, by the effect of the estrogen-progesterone ratio.

The wide variety in dosages and combinations of pills, has led to methodological problems in this research. In well-designed studies, Paige (1971) compared sequential versus combination pills. There was a finding that the negative affect during the menstrual cycle does not occur among women using combination oral contraceptives. Hamburg et al. (1968) report that premenstrual tension is relieved by some oral contraceptives.

Research efforts are underway that have attempted to study the relation of oral contraceptives to symptoms of depression. This work has been studied and reviewed by Kutner and Brown (1972). In an ongoing longitudinal project they have been studying women who have never used *the pill,* past users (those who tried it and have discontinued), and current users. Their population is 8,083 women who are subscribers to the Kaiser Foundation Health Plan in Northern California and willing to cooperate. They found no evidence that use of oral contraceptives aggravated depression history. There was no relation between number of months of pill use and depression history. They found that in current users with a depression history, current measures of depression were independent of dosage of progestin. Their most significant findings related to past users. They showed the highest incidence of distressing premenstrual irritability and a significantly greater history of severe depression during or after pregnancy. These past users seem to constitute a group that deserves intensive endocrinological and behavioral study.

Sex Hormones and Sexual Behavior

Rhesus monkeys are a useful laboratory model for the experimental study of sex hormones and sex behavior. They resemble humans in several ways. They have a 28-day menstrual cycle. In laboratory rhesus monkeys there is absence of estrus-associated periodicity of sexual receptivity. They both show periods of heightened sexual receptivity near the mid-point of the menstrual cycle (human studies: Udry and Morris 1968; and rhesus monkey studies: Michael, 1968). Also, as in human beings, there is a striking factor of individual variation in consistent partner preference as a significant determinant of absolute level of sexual activity. In the experimental procedures animals were selected for high levels of sexual interaction.

Luttge (1971) has done an extensive literature review of the role of gonadal hormones in sexual behavior of rhesus monkey and man. Michael has been a major contributor to research on rhesus monkeys. His findings can be summarized as follows. Mounting attempts on the part of the male increase in number towards mid-cycle and ejaculation time becomes shorter. There is a decline of this behavior during the luteal phase. Males spend more time grooming females near mid-cycle than near menstruation. The female rhesus is more agonistic to the male at mid-cycle when sexually receptive. She is less agonistic and more liable to be attacked by the male just prior to menstruation. Female invitations to encourage male mounting behavior seem to remain at a fairly constant level throughout the menstrual cycle but the female success ratio or proportion of such invitations accepted by the male, markedly decline during the luteal phase. There is also a female refusal of male mounting efforts that appears during the luteal phase (Michael 1971).

Using ovariectomized animals and controlled injections of estrogen and progesterone, some mechanisms have been elucidated. After ovariectomy, the cyclic sexual behavior of both males and females described above was abolished. When estrogen was administered male interest was restored to normal levels as evidenced by attempted mounts and grooming activity. Administration of progesterone showed that the female refusal of male mounts is due to a progesterone effect. Concommitantly, males lose interest in females receiving high doses of progesterone. Their mounting attempts sharply decline. In human beings, there has been a reported loss of sex drive in couples where the woman has been on progestational oral contraceptives for long periods of time (Grounds 1965, Kane et al. 1967, Grant and Mears 1967).

Michael (1971) administered progestational oral contraceptives to his normal, intact rhesus females using combination pills for six to seven months. It was then changed to a sequential regimen for a similar period of time. Changes in sexual interaction occurred. The most conclusive finding so far has been a progressive decrease in the number of male ejaculations per test with successive cycles of treatment with the combination pills. The male's thrusting activity seems to be impaired by changes in the quality of the vaginal secretion. With the sequential treatment the impairment does not occur (Michael and Plant 1969).

PHEROMONES AND SEXUAL BEHAVIOR

Michael and his colleagues (1968) have shown that olfactory cues mediate sexual initiating behaviors of the male rhesus. They demonstrated an estrogen-dependent vaginal pheromone mechanism. Male monkeys rendered unable to smell, failed to bar-press to gain access to estrogen-treated females after it had been demonstrated that with normal olfaction they consistently did so. An estrogen-stimulated vaginal secretion, *copulin,* has been identified as being the stimulant to the male. There is individual difference in the extent of production of the substance by the female rhesus but there is a general tendency to show increases near mid-cycle.

Copulins (volatile fatty acids) have been identified in the vaginal secretions of human females (Michael et al. 1974). In cross-species experiments these human vaginal secretions were shown to have sex-attractant properties in rhesus monkeys via an olfactory route. The data reveal that humans also show increased production of copulin near mid-cycle. However, no cyclic increase was found and total amounts were lower in women taking contraceptive pills.

It has been postulated that a pheromone mechanism may operate to produce menstrual synchrony in human females (McClintock 1971). It was found that groups of women living together, particularly close friends, tended to have entrainment of menstrual cycles to the same timing. Also, exposure to males was found to affect cycle length. Rossi (1974), in the study previously cited, confirmed that women with high contact with males (seeing men three or more times per week) had an average menstrual cycle of 25.5 days. Women with low contact with males had an average menstrual cycle of 33.6 days. Pheromone mediation is postulated by Rossi for this effect on cycle length.

Menstrual synchrony has been observed in the hamadryas baboon. Kummer (1968) studied the timing of the menstrual cycles of harems of females living in units within an ecological area. He found that the group of females attached to a particular male were synchronized both monthly and seasonally in their estrus. Within each cluster of attached females there was synchronization to each other but there was a notable lack of synchrony across the different troops living in the same area. Kummer, too, postulates a pheromonal explanation.

Although psychosocial factors do seem superordinate in human sexual behavior, it is possible that olfaction may play a significant role. It is known, for example, that pubertal girls and mature women can smell musk substances that sexually mature males cannot perceive unless they have received estrogen injections (Le Magnen 1952). Bruce (1970) and Michael and Keverne (1968) have done the most recent reviews of the role of pheromones and sexual behavior in lower mammals, primates, and humans.

TESTOSTERONE AND FEMALE SEXUAL BEHAVIOR

Recently, there has been interest in the production of testosterone in human females during the normal menstrual cycle and its effects (Ismail et al. 1968). There appears to be a cyclic pattern of testosterone excretion in urine that seems to parallel estrogen production with a peak at mid-cycle (ovulation). This has been replicated by Persky (1974). Most of the testosterone in females is produced by the adrenal cortex.

Testosterone has been linked to sex drive. There are clinical reports that women receiving testosterone for menstrual disorders or malignant disease have rather consistently reported heightened libido (Abel 1945, Herrmann and Adair 1946, Foss 1951, Waxenberg 1963). These reports are confounded, however, by the variables relating to the underlying illness. In human beings, libidinal responses were studied by Bakke (1965) in a double blind design in which women successively received estrogen, estrogen-progestin, and placebo. A significant increase in sex drive and sex activity was reported when the progestin-containing preparation was administered. This effect was postulated to be due to the known androgen-like properties of progestin.

Everitt and Herbert (1969) used the adrenocortical hormone, dexamethasone, to suppress the hormone production of the adrenal cortex of female rhesus monkeys. They found that androgen levels fell and sexual receptivity fell as measured by refusals of male mounting attempts and decline in female presentations, or sexual invitations to the male. The receptivity was restored only with administration of testosterone.

PSYCHOSOCIAL EFFECTS ON TESTOSTERONE LEVELS IN MALES

For many years there has been an effort to control or punish male antisocial behaviors such as rape or incest by manipulation of testosterone levels. Such methods as castration (Bremer 1959), estrogen administration for sex offenders, and more recently administration of androgen-blocking agents such as medroxyprogesterone acetate (Money 1970) and cyproterone acetate (Cooper et al. 1972) have been used. The effects of these procedures have been variable. As reported by Bremer, the most notable reduction of sexual interests after surgical castration occurs in the youngest and eldest patients. In general, the young adult and middle-aged group were reported to be unaffected by the sexual behavioral effects of surgical castration. Laschet and Laschet (1969) found that administering the antiandrogen cyproterone acetate to *hypersexual* men was not successful in reducing abnormal sexuality in those men whose problem was related to central nervous system functioning (for example, brain lesions) and not hormones.

Recently, there have been reports of low testosterone levels in homosexual men (Ismail et al. 1968; Kolodny et al. 1971; Loraine et al. 1971). It is not a universal finding and when it occurs the interpretation is not clear. Recent primate work by Rose et al. (1971) on changes in testosterone levels in rhesus monkeys under situations of stress and sexual stimulation suggest that further human studies must be designed to account for psychosocial factors in the regulation of testosterone levels. Rose found that animals subjected to a subjugation and loss of status in the dominance hierarchy show a drop in testosterone. Male monkeys placed in a situation of sexual stimulation with receptive females show a rise in testosterone. Kreuz et al. (1972) also studied testosterone levels in young men under varying stress in officer training school. They found that plasma testosterone levels were significantly lower during the early, highly stressful period of training. This represents the first evidence in humans supporting the hypothesis that psychological stress suppresses levels of circulating plasma testosterone. It is possible that, in addition to psychic cortical effects, these interactions may be directly linked to the adrenal hormone system. Frankenhaeuser (1972) has shown that males and females have comparable adrenalin secretion under conditions of inactivity and rest. However, when stressed, men secrete significantly more adrenalin than women in the test situation. The work of Ward (1974) in utilizing severe prenatal environmental stress to cause decrease in androgen production in rats has been described (see page 172). She postulates an ACTH elevation by stress which causes a reciprocal decrease in gonadotropin. This, in turn, leads to decrease in androgen secretion by the fetal testis. Rivarola et al. (1966) demonstrated that ACTH stimulation in normal males leads to depression of blood testosterone.

The research on sex hormones and sex behavior has shown that testosterone and estrogen are each capable of paradoxical effects of increasing the normal sexual behavior of the opposite sex in addition to increasing the normal sexual behavior of the same sex animals (Whalen 1974; Clemens 1974). Some of the effects may be due to the interactions of the cross-sex hormones that are normally present. The male hormone testosterone is present in normal females in small amounts and is produced by the ovary and the adrenal gland. The female hormone, estrogen, is present in normal males in small amounts. The estrogen found in males is converted from testosterone (and androstenedione) peripherally in the liver but also in the brain in the preoptic area of the hypothalamus (Naftolin et al. 1971; Reddy et al. 1973). These studies add another complexity to the unravelling of the interplay of factors that affect sex differences in behavior.

AGGRESSION

There is no definition of aggression that is generally agreed upon by either biologists or social scientists. However, to avoid confusion, there is a need to specify my use of the term in this discussion. Human aggression refers to attack or threat behavior, physical or verbal, aimed at physical injury or intimidation of another individual. In my view, this definition does include predatory, defense, and dominance behavior. Undirected motoric or verbal outbursts or accidental injuries to others are not included.

There is a long mammalian heritage of sexual dimorphism in aggressive behavior. In the course of primate evolution, male aggression has been characteristic. Livingstone (1967) discusses the role of men in fighting and warfare as a selection pressure. Freedman (1967) has documented a prevalence and consistency of aggressiveness over the course of human history that is far greater than is generally appreciated. There are coadaptations of anatomical structures, physiology, and temperament for the expression of human male aggression that have been described (page 161). Structural and behavioral signals that serve to turn off aggression between individuals have also shown differential selection by sexes. Females are high in these attributes (page 162). The sexual dimorphism of aggressive behavior, although it is reliably established and a major variable defining masculine and feminine behaviors, is not exclusive. Female aggression does exist and can be intense, particularly in defense of young. The frequency and modal level of female aggressive behavior is much lower than that of males.

It is likely that the costs of high aggression, in time spent in agonistic encounters, risk of disabling injury or death, and possibility of fatal aggression towards offspring, probably had significant selective disadvantage for high aggression in ancient human females who had the care of children with a lengthy period of immaturity and dependence. For early hominid males, on the other hand, high aggressiveness probably enhanced potentiality for reproductive success despite the costs of aggression. Highly aggressive males (possessing the appropriate physical and physiological concomitants) were more likely to be successful in efforts to defend against predators, to obtain protein food (hunting of large game), to control resources in times of scarcity, and to have access to preferred females.

In primates, testosterone mediates aggressive response. The role of testosterone in programming the fetal brain for sex differentiated aggressive responses has been established for many mammals and for human and nonhuman primates. In primates, the infancy and childhood effects are expressed in sex differentiated juvenile play patterns (Money and Ehrhardt 1972, Ehrhardt and Baker 1974; Phoenix et al. 1968). For nonhuman primates aggressive behaviors of hitting, threat, and intimidation have been well described (Dolhinow and Bishop 1972; Rosenblum 1961; Hall and DeVore 1965). High male aggression in human children is also well documented (Oetzel 1966, Feshbach 1970; Maccoby and Jacklin 1974).

Circulatory levels of testosterone can be correlated with level of aggressive behavior. The loss of aggressiveness associated with prepubertal castration can be restored by replacement testosterone therapy. Using college students, Persky et al. (1971) found a high correlation between plasma testosterone and scores on a hostility scale. Other investigators using comparable subjects and scales (Doering 1971; Meyer-Bahlburg 1974), and Kreuz and Rose (1972) using a prison population, were unable to replicate Persky's findings. In the prison population, there was, however, a significant correlation between the age of first conviction of a violent crime and testosterone level. This suggests that definitive answers will depend on a research strategy that relates testosterone level to the *behavior* of an individual in a situation where the psychosocial variables are controlled rather than depending on the use of questionnaires and personality inventories. This is particularly important since the effect of psychosocial situation as a determinant of testosterone level is now known.

The biological basis of male aggression in mammals is generally recognized. However, in studies of human aggression, understanding the role of social learning is also recognized as the major challenge. Social learning offers the best hope for exerting effective controls on and fostering constructive channeling of, aggressive tendencies. Despite the complexities introduced by the human capacity for language and symbolic thinking, animal studies may reveal some generalizations that have heuristic value for human studies and give some clues to factors that help to shape social learning. It seems justified to examine patterns of behavior that have characterized animals who are closely related to us phylogenetically and who show comparable evolutionary adaptations. In this endeavor, as in cross-cultural surveys, the emphasis must be on deriving the structural principles rather than focusing on species-specific or culture-specific details.

Some of these generalizations are as follows:

1. Appearance of male aggression or distinctive male precursor juvenile play patterns (rough-and-tumble play) early in development with little learning necessary.

2. Preference for same-sex playmates with heightening of activity levels, threat behavior, and mock-fighting in male-male pairs or groups. This pattern has been characteristic of most primate groups (DeVore 1965; Dolhinow and Bishop 1972). In human beings this is described by Blurton-Jones (1972) and many confirming studies are summarized by Maccoby and Jacklin (1974). This has implications for social learning and the practice and reinforcement of aggressive behaviors preferentially for males.

3. Sex difference in observational learning preference. This type of learning is characteristic of primates. In monkeys and apes, juvenile males are attracted to adult males and juvenile females are attracted to mothers and infants (Hall and DeVore 1965; van Lawick-Goodall 1968; Fossey 1972). The same phenomenon was classically demonstrated by Maccoby and Wilson (1957) in a film content memory experiment that showed that boys preferentially recalled aggressive content when enacted by boys, and girls selectively recalled social interaction when enacted by girls. Bandura et al. (1961, 1963) also found that compared to girls, boys showed greater imitation of physical and verbal aggression when the adult model was male. Experiments show that even if girls are exposed to learning aggressive content, they are less apt than boys to spontaneously display the learned behavior. If reward is offered, girls can and will do so (Bandura 1965).

4. Sex difference in serving as stimulus for eliciting aggressive behavior. Moyer (1974) cites more than 20 examples across mammalian species in which inter-male aggression is very high and there is concurrent tendency to avoid attack of females. In a nursery school study of reinforcement of aggressive behaviors, Patterson et al. (1967) found that children who responded to attack by yielding or crying were chosen for repeated attack by males. Girls were significantly underchosen for this repeat attack despite the fact that there was just as much yielding and crying by girls. In adults,

187

it was shown by Taylor and Epstein (1967) that even when females were highly provocative, males preferentially refrained from retaliation against women in an experiment involving delivery of electric shocks.

5. Signals of submission that turn off aggression. The ethological literature has many descriptions of the anatomical and behavioral characteristics that render an individual less likely to elicit aggression. It also describes other species-specific signals that can be actively employed by a subjugated animal to turn off continuing attack and prevent serious injury. Guthrie (1970) has discussed the evolution of these signals in human beings. He does not deal with verbal cues. There has been little attention to this area in child development. Krebs (cited by Hinde 1974:132) found that low ranking nursery school children tend to smile more than dominant children on initiating social contacts. He views this as appeasement.

6. Specific stimulus of *stranger* males of same species (conspecifics). For most of the species in which male aggression occurs (Moyer 1974) the entry of a same species stranger male into an established group living situation will elicit immediate aggression regardless of the context of the situation or the demeanor of the stranger. This behavior is very prominent in primates. It has been reported in human beings even in groups regarded as unusually peaceable such as !Ko-bushmen (Eibl-Eibesfeldt 1974).

The evolution of human aggression occurred when human societies were small social groups and the individuals were all known to each other. With the change to a modern, industrial world characterized by urbanization, mobility, and large and shifting social groups, there is no reason to assume that these adaptations, derived from an earlier time, have their same general utility or adaptedness.

For many careers and occupations, the level of social skill in dealing with people may be much more important than aggressiveness in achieving success in contemporary society. We need to know as much as possible about the mechanisms underlying the learning and display of aggression for both sexes so that our efforts at shaping social learning for the modern world can take into account the learning preferences and tendencies that have evolved under quite different conditions of life.

SEX DIFFERENCE IN THE LATERALIZATION OF THE BRAIN

The functional division of the human brain into a dominant hemisphere and a nondominant hemisphere appears to be an evolutionary change that is related to the development of language and also to the adaptive advantage of *handedness* in toolmaking and other skills of manual dexterity. Because of the crossing over of nerve pathways, the dominant hemisphere is contralateral to the dominant hand. Since most people are right-handed, the left hemisphere is usually dominant. The speech and language centers are located in the dominant hemisphere. Spatial ability has been shown to be localized in the minor (nondominant) hemisphere.

Our information concerning lateral specialization of the brain comes chiefly from patients who have asymmetric lesions of the hemispheres. A great deal has also been learned through surgical interventions necessary for tumors and epilepsy. Of late, new approaches have been possible through the technique of commissurotomy, which is the surgical disconnection of the cerebral hemispheres accomplished by cutting the corpus callosum. (Sperry and Levy 1970). Intractable epilepsy has been found to be improved by this procedure.

Witelson and Pallie (1973) measured cerebral hemispheres of normal adult and newborn brains (deaths due to nonneurological causes). In the neonates, significant sex difference was found. Male neonates showed little anatomical difference between the hemispheres. In the females, the superior temporal surfaces of the left hemisphere were significantly larger than the right. In adults, the left or dominant hemisphere was consistently larger than the right for both sexes. Wada (cited by Milner 1974) reported that asymmetry of hemispheres with a substantially larger planum temporale on the left can be demonstrated as early as fetal age of 20 weeks.

There is a well-documented superiority of girls in verbal skills by ten years of age (Jacklin and Maccoby 1972, McCarthy 1954, Tyler 1963). Available evidence strongly suggests that this difference may be due, in part, to the way in which the brain is organized. At the same time, males have been shown to be more proficient in spatial tasks than females. Buffery and Gray (1972) have given an excellent, detailed review of the present state of evidence for sex differences in linguistic and spatial abilities and the research on the asymmetries of cerebral function. The following description covers the main lines of evidence and also offers some interpretations from an evolutionary perspective.

Language

The lateralization of cerebral language function has been studied in normal subjects through the techniques of dichotic listening, (Kimura 1967; Knox and Kimura 1970). Dichotic presentation refers to the simultaneous presentation of different stimuli to the receptor organs (eyes or ears). Normally, in such simultaneous presentation, the stimuli are not processed by the two ears or eyes with equal proficiency. These perceptual asymmetries seem to reflect the normal functional asymmetry of the brain.

Using dichotic auditory methods, superiority of the right ear (and, therefore, left hemisphere) for the perception of words and speech sounds (verbal stimuli) has been demonstrated for mature individuals. With dichotic presentation, environmental (nonverbal) sounds are more accurately identified by the left ear (right hemisphere). This work demonstrates a specialization and functional differentiation of the hemispheres along the verbal/nonverbal dimension. When children were studied at five years of age, young males were distinctly more proficient in the processing of nonverbal stimuli (Knox and Kimura 1970). The authors concluded that in their sample of right-handed children "The right and left hemispheres had begun to show a functional differentiation along the verbal/non-verbal dimension by age five" (p. 234). The superiority of girls on the verbal identification is attributed by the authors to differential maturation, with the girls maturing faster.

There is some supportive evidence for faster rate of maturation of the lateralization of the left hemisphere in girls from a totally different line of inquiry. Taylor (1969) studied 158 cases of temporal lobe epilepsy in terms of the sex distribution of the age of onset of the first seizure. There is a well-established hypothesis that a potential seizure-producing insult usually affects the less functionally active hemisphere. Taylor found that there was a consistently greater proportion of boys than girls who were continuing to be affected by left-sided seizures up until five years of age. In girls, the left-sided seizures sharply declined after two years of age. Right-sided lesions were equally prevalent for both sexes. This would seem to point to earlier maturation of lateralization of the left hemisphere.

Earlier maturation of language centers in the female may contribute to high verbal facility in girls through predisposition to early vocalizations and vocal responsiveness to human contacts. An attentiveness to and preference for verbal interaction may develop that enhances language development. Sherman (1971) has a similar hypothesis but also posits a competitiveness between verbal and spatial cognitive styles that is as yet unproven. A greater predictive power of early infant babbling and later I.Q. for girls than boys appears to exist (Cameron 1967; Moore 1967). The evolutionary advantages of language are probably equally great for males and females.

Although there have been several negative studies (Maccoby and Jacklin 1974), greater verbal stimulation of girls than boys in mother-infant interactions has been described (Lewis and Goldberg 1969; Moss 1967). If these findings prove to be accurate and girls do tend to receive more verbal stimulation than boys, this social interaction would then potentiate the maturational advantage and further enhance the verbal skills of girls.

The earlier lateralization of the cerebral functioning in females reflects a prenatally organized sex difference in rate of maturation. Early maturation of females is detectable from 20 weeks of fetal age through adolescence. For example, the ossification of the skeletal centers in the female is four weeks ahead of the male at birth and has advanced to an 18-month difference by puberty. The maturational lag of the male is also dramatically reflected in the timing of puberty, which occurs, on average, two years later than in the female (Tanner 1960; McCammon 1962; Christie et al. 1950; Hansman and Maresh 1961).

Spatial Perception

Much valuable information about the specialized functions of the minor hemisphere has also been gained through the commissurotomy technique. Sperry and Levy (1970) have found that knowledge about minor hemisphere function can only be gained through special tests that utilize nonverbal forms of motor expression. Inasmuch as language is located in the dominant (left) hemisphere, information can be elicited about function of that hemisphere through verbal report of the commissurotomy patient. This verbal report is not possible with the right or *mute,* minor hemisphere patient. Through their studies, Sperry and Levy have concluded that the minor hemisphere

> is indeed a conscious system in its own right . . . perceiving, feeling, thinking and remembering at a characteristically human level. More than this it has been shown that the minor hemisphere is distinctly superior to the leading hemisphere in the performance of certain types of tasks . . . as for example, in copying geometric figures, in drawing spatial representations and in assembling Kohs blocks in block design test.

Nebes (1974) studied commissurotomized patients to determine the extent of the functions of the minor hemisphere. He found that the right or minor hemisphere has the specialized ability to generate a holistic perception from parts or fragments. This function is very important as it provides organization to an environment that often presents incomplete or limited data to the senses.

Spatial localization in the brain has also been studied through dichotic techniques. Kimura (1969) demonstrated that a point can be more accurately located when presented to the left visual field than when presented to a corresponding position in the right visual field. It was also demonstrated that "on the localization task employing location within squares, the left field (spatial) superiority was absent or less marked for females compared with males" (p. 456). McGlone and Davidson (1973) studied right- and left-handed men and women on a dichotic words test, visuospatial tasks, and tachistoscopic dot enumeration task. Men performed better on the spatial tasks. Men also showed more right hemisphere dependence.

Other investigators have also found male superiority in spatial ability using other perceptual tasks. Witkin et al. (1954, 1962) found differences in three tests of spatial discrimination: the embedded figure test, rod and frame test, and the tilting room test. After age eight years, males did consistently better on all of these tests. Prior to that age, results were variable (Coates 1974). Porteus (1965) has found that his maze test of spatial ability has yielded consistent results of male spatial superiority across a large number and wide range of cultures (from Australian aborigines to European and American school children). Finally, there is strong support for the presumption that genetic factors participate in determining spatial ability because in girls with Turner's syndrome, a 45, XO condition, there is a profound deficit in spatial perception (Alexander et al. 1966). Levy and Nagylaki (1972) have proposed a genetic model for the inheritance of spatial ability and handedness. They postulate that both are recessive alleles of a sex-linked gene. Two other recessive alleles controlling hemisphere dominance are also postulated for the same gene. These four alleles can assort independently.

Sperry (1974) also reports the tendency for language to develop at the expense of competing nonverbal functions. Milner (1974), in summarizing recent findings in the area of hemispheric specialization, points out that most tasks require the combined and complementary efforts of both hemispheres for efficient solutions. For example, even where the right hemisphere has a major function in a spatial perception task, there is normally participation of the left hemisphere in supplying a useful label.

From an evolutionary perspective there would appear to be adaptive advantage in a tendency for male superiority in spatial skills. Washburn (1969) has pointed out that most monkeys spend their lives in an area of two or three square miles. The gorilla and chimpanzee have a range of roughly 15 square miles. Washburn indicated that one of the remarkable characteristics of human beings is that even the most primitive of men operate over hundreds of square miles rather than these small areas. Improved spatial abilities may have played a role in facilitating the great increase of human territorial range. These abilities would also have given an adaptive advantage in hunting and warfare. One of the skills shared by man and chimpanzee is the ability for accurate, powerful, aimed throwing. Van Lawick-Goodall (1968) reports that the throwing of stones and other objects is done exclusively by males and in the context of agonistic encounters. In humans, the throwing of spears has a long history in both hunting and warfare. The importance of human male throwing is underscored by the adaptations of the arm and shoulder girdle, which make powerful overhand throwing possible. The anatomy of the woman predisposes her to throw underhand (Washburn and McCown 1974).

Affiliative Behaviors

Maccoby and Masters (1970) have done a comprehensive review of the status of dependency research. The concept of dependency grows out of social learning theory (Sears 1963). As used in that literature, dependency has included all contact-seeking, help-seeking, approval-seeking, and attention-getting behaviors. It was hypothesized that these behaviors should be positively correlated with each other. This has not proven to be the case. Maccoby and Masters (1970) noted that proximity-seeking and attention-getting are separate clusters of behavior with differing antecedents and outcomes. Data have been reanalyzed by them using categories based on findings from cluster analysis. Only negligible sex differences in dependency are found but the data shows many gaps on issues of nurturance and empathy. Maccoby and Jacklin (1974) also find that the tendency for a child to relate to other children is independent of the tendency to relate to adults.

For the reasons of lack of consistency or predictability of dependency behaviors cited above and other reasons to be outlined, the concept of attachment would appear to offer advantages in giving direction to future research into socialization across the life cycle. Ainsworth (1972) has systematically compared concepts of attachment and dependency and also cites much of her own work in the field of attachment. The concept of attachment was introduced by John Bowlby in 1958. Since that time, it has become a topic of major research interest.

In brief, the concept of attachment derives from research in ethology and embodies an evolutionary perspective. It postulates a genetic predisposition to become attached to specific individuals under appropriate conditions of reciprocal stimulation. It has been usually described in the context of mother-infant interactions. Yarrow and Pedersen (1972) emphasize the strong affect associated with the interpersonal relationship. To them, it is also the prototype for later relationships with significant others. It seems worthwhile to review the concept of attachment in a broader context than merely a mother-infant relationship and see how attachment or social-bonding behavior, as it can also be designated, might have played a significant role at all ages. In the course of this, I should like to focus on ways in which attachment behaviors may have come to show a sexual dimorphism in adult individuals.

Hinde (1974) states that "over the animal kingdom as a whole, responsiveness specific to particular individuals is characteristic of only a small minority of species" (p. 16). As has been discussed previously, in human societies a premium is placed on social organization and interpersonal relations that foster long-enduring social bonds between adults and the vulnerable child. As a corollary of this, adult male-female bonding in the service of food-sharing and protection becomes important. A premium would have been placed on the capacity of the male to form attachments to the young of the consort female (paternalism). Mention has been made of the high degree of cooperation among males that appears to have characterized male hunting and warrior groups in the early human being. It would appear, therefore, that biosocial mechanisms to assure firm and enduring social bonds (attachment) between specific group members at all ages were major evolutionary adaptations.

Mother-Infant Attachment

Mother-infant attachment is seen as fulfilling many functions for the child: nurturance, protection from predators or accidents, and maintenance of an appropriate level of stimulation for the infant. Fear of strangers and amount of exploratory activity in infants are used to judge infant attachment. Heavy emphasis is placed on proximity-seeking and/or contact behavior in relation to the attached person. For the child, the degree of contact varies in type and intensity in relation to the situation. At times of fear, close physical contact is demanded. During play, intermittent, fleeting, and visual or vocal contact may suffice. Signals from the infant that elicit attachment behaviors on the part of the mother are sucking, clinging, crying, smiling, babbling, and, later, calling to and ambulatory following of the mother.

It is postulated that across primate evolution, females who have had a strong propensity to respond in a nurturant way to these infant signals have had a strong selective advantage. The human infant has been shown to respond selectively to auditory (Eimas 1971; Eisenberg 1969; Hutt et al. 1968; Kagan 1971) and visual stimuli (Haynes et al. 1965; Kagan and Lewis 1965) in those ranges that are characteristic of human beings. Human beings seem to have an ease of learning of attachment to a specific individual.

There is a clear primate heritage for this. The mother and her children have been well described as a stable and enduring social unit over many years in higher primates (DeVore 1965; van Lawick-Goodall 1968). As has been mentioned, attachment depends on an interaction of individuals under appropriate conditions of reciprocal stimulation. There would, therefore, also be a clear advantage to all young infants, both male and female, to emit the appropriate stimuli which will elicit attachment behavior on the part of the mother and to respond equally to her reciprocal stimulation. One would not expect and research does not demonstrate significant sex differences in attachment behaviors displayed by young infants (Schaffer and Emerson 1964; Ainsworth 1972; Lewis and Weinraub 1974).

The strength of attachment is typically measured in terms of reactions of distress engendered in the infant by separation from the mother. Less has been written about the very highly charged emotional reactions engendered within the mother if she is separated from her child. Female aggression is most reliably noted in defense of her child. Mothers also show proximity-seeking behavior when separated from the infant. These maternal responses are evidences of *her* attachment to the infant.

The libidinal aspects of maternalism have been reviewed by Niles Newton (1973). She describes the maternal enjoyment and gratification associated with lactation and nursing and, at times, childbirth. Some comparisons are made with sexual gratification. There is also hormonal priming of post-partum mothers that facilitates maternal behavior and lactation. Preliminary work in rats suggests that maternal behavior is influenced by gonadal effects on the developing brain. Genetic males who were feminized very early by antiandrogens and administered female sex hormones as adults, showed licking, pup-retrieval behavior, and lactation in response to newborn rat pups (Neumann, Steinbeck and Hahn 1970).

Paternalism

There has been some research into the conditions that elicit paternal behavior in animals (Rosenblatt 1967). Proximity is important. Care of the young could be elicited from normal adult male rats after prolonged contact with the litter (five to seven days). Paternal care consisted of licking and retrieving young to nest. Male adoption of orphaned infants has been reported in some nonhuman primates. In reviewing primate literature, Mitchell (1969) finds that in monogamous or one-male primate groups, paternalism is especially high.

Aggression is distancing behavior and as such is antithetical to attachment. The fact of higher aggression and dominance behavior in males and between males has been generally acknowledged. Early infancy is an exception. There are no consistent sex differences in measures of activity level or other variables that might relate to aggressiveness in neonates (Korner 1973). In fact, infants and children provide the strongest structural and behavioral cues for turning off aggression. It is likely that this is important in providing the conditions under which paternalism or attachment to specific young can develop. The distancing effect of aggression is minimized and the adult male and infant can freely and frequently interact in the mutually reinforcing reciprocal behaviors that lead to attachment.

Male-Female Pair Bonds

The criteria used for mother-infant attachment can equally well be applied to the measurement of the strength of adult male-female bonds. In both instances, there is intense gratification associated with behaviors that maintain bonds and severe emotional distress and proximity-seeking when separation occurs or is threatened.

The lack of estrus in the human female represents an uncoupling of the ovulatory and copulatory cycles. The explanation does not lie in an increased likelihood for fertile matings. On the contrary, the probable loss of fertility must be outweighed by other gains. Sexuality is freed to become the vehicle of powerful inter-individual attraction and the basis for enduring affectional pair bonds (Beach 1975). The adult male-female attachment is further enhanced by the adaptation of maternal affectional behaviors to the adult heterosexual relationship.

Dimorphism in Attachment Behavior

Primate studies shed some light on an aspect of the post-infancy sex differences in attachment behavior. Jensen (1968) reports that in pigtailed macaques under three months of age, there was no difference in attachment behavior. After three months, the males pulled away from and hit the mothers more. Mothers were more punitive and distant from older male infants. Mothers touched and groomed older female infants more. Rosenblum (1974) found no attachment differences in bonnet macaques until six months of age. After that age, females maintained close relationship with mother and the social group. Males were more active, explored more, and left mother more.

Yarrow and Pedersen (1972) have shown in human beings that there is a positive relationship between the degree of matching of mother and child in temperament, activity level, and modality preferences and attachment. The testosterone-mediated sex differences in temperament and activity levels of male juveniles as a factor in distancing mother-child interaction is highlighted in these nonhuman primate studies.

Attachment versus Dependency

The prevalent use of the terms dependency or passive-dependent tendencies to characterize females seems less useful than an attachment concept broadly defined, as outlined above. Dependence, as applied to an adult, is a pejorative term with implications of immaturity and helplessness. Beyond earliest infancy, attachment does not have those connotations. Even looking at later infants, their exploratory and more *independent* behavior is positively correlated with secure attachment to a maternal figure (Ainsworth 1972). Viewed from the perspective of the mother, the *attached* mother is far from passive in her behavior towards an infant or child of any age.

Maccoby and Jacklin (1974) review the evidence for humans concerning sex difference in patterns of social approach. They cite the Whiting and Pope six cultures studies. In five out of six cultures, girls scored higher than boys on nonaggressive touching and holding behaviors. They also cite studies that show that girls tend to focus in play on one or two best friends. Boys tend to form game- or skills-oriented groups rather than groups based on "chumship." These sex-differentiated play preferences were confirmed by Omark (cited by Maccoby and Jacklin 1974) in a cross-cultural study of American, Swiss, and African children. Studies are cited that show that for girls, but not for boys, proximity is a sensitive indicator of degree of friendship as measured by projective test and direct observation. Hinde (1974) presents a detailed discussion of the nonrandom spatial dispersion of individuals in primate troops as a function of social relationship. He makes a general statement that "agonistic interactions promote dispersion and affinities promote cohesion" (p. 355). Play of boys is very prone to be aggressive, and it is likely that this tends to promote the dispersion in social relationships that was noted by Maccoby and Jacklin in their review. Adult females do seem to show a greater preference and capacity for close, enduring relationships with significant others.

Attachment seems a more useful concept also because it is a term associated with emotional health and normality. For human beings, the inability to form attachments is the pathological condition. Therefore, the imputing of pathology to descriptions of modal behavior of women would be avoided.

Female dependency has been perceived as a *need* for the company of others, their attention and approval. This view seems to represent a partial perception that overlooks the reciprocal aspects of an attachment relationship.

CONCLUSION

The approach taken in this paper is biosocial, or that of behavioral science. As yet there is not a generally accepted term that embraces a wide variety of knowledge, understandings that have been the properties of different departments of the university. But from a practical point of view, the understanding of sexual dimorphism depends on evidence from: evolution, animal behavior, genetics, hormones, embryology, and development. The all-important brain matures in a biochemical and social environment that may be different for males and females. In every culture the rewards and costs have been different for males and females. But whatever insights we may gain from all this information, we must not forget that the opportunities of the modern world are really new. The tasks which used to require the strength of males are now performed by machines. Women no longer need spend most of their adult life producing babies. Yet in this new and rapidly changing world people still form attachments. There are the young and the old. There is still the problem of the good life, and that life is made up of the lives of males and females who need to be understood by every method that can be brought to bear on the understanding of us.

Acknowledgments

This review is an outgrowth of a workshop on the Biological and Cultural Bases of Sex Role Difference, sponsored by NIMH, Division of Extramural Research. It was organized by the author and took place in Palo Alto, June 1972. The work was also supported by the Commonwealth Fund.

I wish to express my deep appreciation to all of the participants and to Betty Pickett and Lorraine Torres of NIMH. I also would like to express my deep appreciation to David Hamburg, Sherwood Washburn, and Frank Beach for their stimulation, criticism, and encouragement. The research and editorial assistance of Barbara Miller, M.S.W., in the writing of this manuscript is gratefully acknowledged.

References Cited

Abel, S.
1945 Androgenic therapy in malignant disease of the female genitalia. Amer. J. Obstet. and Gynecol. 49:327–342.

Ainsworth, M. D.
1972 Attachment and dependency: a comparison. In Attachment and Dependency, J. L. Gerwitz, ed. New York: V. H. Winston and Sons.

Alexander, D., A. Ehrhardt, and J. Money
1966 Defective figure drawing, geometric and human in Turner's syndrome. J. Nerv. and Mental Disease 142:161–167.

Altmann, M., E. Knowles, and H. Bull
1941 A psychosomatic study of the sex cycle in women. Psychosomat. Med. 3:199–225.

Andersen, H.
1966 The influence of hormones on human development. In Human Development. F. Faulkner, ed. Philadelphia: Saunders.

Arai, Y., and R. Gorski
    1968 Critical exposure time for androgenization of the rat hypothalamus deter-
        mined by antiandrogen injection. Proceedings of the Soc. for Exper. Biol.
        and Med. 127:590–593.

Bakke, J. L.
    1965 A double-blind study of progestin-estrogen combination in the management
        of the menopause. Pacif. Med. and Surgery, 73:200–205.

Bandura, A.
    1965 Influence of models' reinforcement contingencies on the acquisition of imita-
        tive responses. J. of Pers. and Soc. Psychol. 1:589–595.

Bandura, A., D. Ross, and S. Ross
    1961 Transmission of aggression through imitation of aggressive models. J. Ab-
        norm. Soc. Psychol. 63:311–318.

    1963 Imitation of film mediated aggressive models. J. Abnorm. Soc. Psychol.
        66:3–11.

Bardwick, J.
    1971 The Psychology of Women. New York: Harper and Row.

Beach, F. A.
    1947 A review of physiological and psychological studies of sexual behavior in
        mammals. Physiol. Rev. 27:240–307.

    1968 Factors involved in the control of mounting behavior by female mammals. In
        Perspectives in Reproduction and Sexual Behavior. M. Diamond, ed.
        Bloomington: Indiana University Press.

Beach, F. A., R. G. Noble, and R. K. Orndoff
    1969 Effects of perinatal androgen treatment on responses of male rats to gonadal
        hormones in adulthood. J. Compar. Physiol. Psychol. 68:490–497.

Benedek, T., and B. Rubenstein
    1939a The correlations between ovarian activity and psychodynamic processes. The
        ovulation phase. Psychosom. Med. 1:245–270.

    1939b The menstrual phase. Psychosom. Med. 1:461–485.

Berndt, R. M., and C. H. Berndt
    1964 The World of the First Australians. Chicago: University of Chicago Press.

Biller, H. B.
    1971 Father, Child, and Sex Role. Lexington, Mass.: D. C. Heath.

Blurton-Jones, N. G.
    1972 Categories of child-child interactions. In Ethnological Studies of Child Be-
        haviour, N. G. Blurton-Jones, ed. Cambridge, Mass.: Cambridge Univer-
        sity Press.

Bowlby, J.
    1958 The nature of the child's tie to his mother. Internat. J. of Psychoanal.
        39:350–373.

Bremer, J.
    1959 Asexualization. New York: MacMillan.

Broverman, I. K., S. R. Vogel, D. M. Broverman, F. E. Clarkson, and P. S. Rosen-
krantz
    1972 Sex role stereotypes: a current appraisal. J. of Social Issues 28:59–77.

Bruce, H. M.
1961 Time relations in the pregnancy-block induced in mice by strange males. J. Reprod. Fert. 2:138–142.

1970 Pheromones. Brit. Med. Bull. 26:10–13.

Buffery, A. W., and J. A. Gray
1972 Sex differences in the development of spatial and linguistic skills. *In* Gender Differences: Their Ontogeny and Significance, C. Ounsted and D. Taylor, eds. Pp. 123–158. Baltimore: William and Wilkins.

Cameron, J., N. Livson, and N. Bayley
1967 Infant vocalizations and their relationship to mature intelligence. Science 157:331–333.

Childs, B.
1965 Genetic origin of some sex differences between human beings. Pediatrics 35:798–812.

Christie, A., M. Martin, E. L. Williams, G. Hudson, and J. C. Lanier
1950 The estimation of fetal maturity by roentgen studies of osseus development. Amer. J. Obstet. Gynecol. 60:133–139.

Clemens, L. G.
1972 Fetal hormones and the brain: behavior dimorphism. *In* Man and Woman, Boy and Girl, J. Money and A. A. Ehrhardt, eds. Baltimore: Johns Hopkins University Press.

1974 Neurohormonal control of male sexual behavior. *In* Advances in Behavioral Biology, Vol. 11: Reproductive Behavior, W. Montagna and W. A. Sadler eds. Pp. 23–53. New York: Plenum Press.

Clemetson, C. A., L. M. Blair, and D. H. Reed
1962 Estrogen and capillary strength. Amer. J. Obstet. Gynecol. 83:1261–1268.

Coates, S.
1974 Sex differences in field independence among pre-school children. *In* Sex Differences in Behavior, R. Friedman, R. Richart, and R, Vande Wiele, eds. New York: Wiley.

Cooper, A. J., A. A. Ismail, A. L. Phanjoo, et al.
1972 Antiandrogen (cyproterone acetate) therapy in deviant hypersexuality. Brit. J. Psychiat. 120:59–63.

Dalton, K.
1959 Menstruation and acute psychiatric illnesses. Brit. Med. J. 1:148–149.

1960 Menstruation and accidents. Brit. Med. J. 2:1425–1426.

1961 Menstruation and crime. Brit. Med. J. 2:1752–1753.

1964 The Premenstrual Syndrome. Springfield, Ill.: Charles C. Thomas.

1967 The influence of menstruation on glaucoma. Brit. J. Opthma. 51:692–695.

Damon, A., H. W. Stoudt, and R. A. MacFarland
1966 The Human Body in Equipment Design. Cambridge, Mass.: Harvard University Press.

D'Andrade, R. G.
   1966 Sex differences and cultural institutions. *In* The Development of Sex Differences, E. E. Maccoby, ed. Pp. 174–204. Stanford, Ca.: Stanford University Press.

DeVore, I.
   1965 Primate Behavior: Field Studies of Monkeys and Apes. New York: Holt, Rinehart and Winston.

DeVore, I., and S. L. Washburn
   1963 A comparison of the ecology and behavior of monkeys and apes. *In* Classification and Human Evolution, S. L. Washburn, ed. Chicago: Aldine Publishing Company.

Dobzhansky, T.
   1969 The pattern of human evolution. *In* The Uniqueness of Man, J. Rolansky, ed. Amsterdam: North Holland.

Doering, C. H., H. K. H. Brodie, H. C. Kraemer, H. Becker, and D. A. Hamburg
   1974 Plasma testosterone levels and psychologic measures in men over a 2-month period. *In* Sex Differences in Behavior, R. Friedman, R. Richart, and R. Vande Wiele, eds. New York: Wiley.

Dolhinow, P., and N. Bishop
   1972 The development of motor skills and social relationships among primates through play. *In* Primate Patterns, P. Dolhinow, ed. New York: Holt, Rinehart and Winston.

Edwards, D. A.
   1969 Early androgen stimulation and aggressive behavior in male and female mice. Physiology and Behavior 4:333–338.

Ehrhardt, A. A., N. Greenberg, and J. Money
   1970 Female gender identity and absence of fetal hormones: Turner's syndrome. Johns Hopkins Medical J. 126:237–248.

Ehrhardt, A. A., and S. Baker
   1973 Hormonal aberrations and their implications for the understanding of normal sex differentiation. Presented at the Society for Research in Child Development, Philadelphia, March, 1973.

   1974 Fetal androgens, human central nervous system differentiation and behavior sex differences. *In* Sex Differences in Behavior, R. Friedman, R. Richart, and R. VandeWiele, eds. Pp. 33–51. New York: Wiley.

Eibl-Eibesfeldt, I.
   1974 Aggression in the !Ko bushmen. *In* 1974 Proceedings of the Association for Research in Nervous and Mental Disease, Vol. 52, S. H. Frazier, ed. Pp. 1–20. Baltimore: Williams and Wilkins Co.

Eimas, P. D., E. R. Signeland, P. Jusczyk, and J. Vigorito
   1971 Speech perception in infants. Science 171:303–306.

Eisenberg, R. A.
   1969 Auditory behaviour in the human neonate: functional properties of sound and their ontogenetic implications. Int. Audiol. 8:34–45.

Everitt, B. J., and J. Herbert
1969 Adrenal glands and sexual receptivity in female rhesus monkeys. Nature 222:1065–1066.

Federman, D. D.
1967 Abnormal Sexual Development. Philadelphia: Saunders.

Ferguson-Smith, M. A.
1966 Sex chromatin, Kleinfelter's syndrome, and mental deficiency. *In* The Sex Chromatin, K. L. Moore, ed. Pp. 277–315. Philadelphia: Saunders.

Ferris, B. G., and C. W. Smith
1953 Maximum breathing capacity and vital capacity in female children and adolescents. Pediatrics 12:341–352.

Feshbach, S.
1970 Aggression. *In* Carmichael's Manual of Child Psychology. Third edition, P. Mussen, ed. New York: Wiley.

Foss, G. L.
1951 The influence of androgens on sexuality in women. Lancet 1:667–670.

Fossey, D.
1972 Physical and behavioral developments amongst individuals of a troop of mountain gorillas (in preparation). *In* Biological Bases of Human Social Behavior, R. Hinde, ed. 1974. New York: McGraw Hill.

Frank, R. T.
1931 The hormonal causes of premenstrual tension. Arch. Neurol. and Psychiat. 26:1053–1057.

Frankenhaeuser, M.
1972 Sex differences in reactions to psychosocial stressors and psychoactive drugs. Report from the Psychological Laboratory, Univ. of Stockholm, No. 367, pp. 1–12.

Freedman, D.
1967 Human aggression in anthropological perspective. *In* The Natural History of Aggression, J. D. Carthy and F. J. Ebling, eds. New York: Academic Press.

Friedman, R., R. Richart, and R. Vande Wiele, eds.
1974 Sex Differences in Behavior. New York: Wiley.

Ginzberg, E.
1971 Introduction. *In* Women and Work, by R. W. Smuts. Pp. V–XVI. New York: Schocken Books.

Glass, G. S., G. R. Heninger, M. Lansky, and K. Talen
1971 Psychiatric emergency related to the menstrual cycle. Amer. J. Psychiat. 128:705–711.

Gorlin, R. J., R. S. Redman, and B. L. Shapiro
1965 Effect of X-chromosome aneuploidy on jaw growth. J. Dental Research 44 (Supp):269–282.

Gorski, R.
1971 Gonadal hormones and the perinatal development of neuroendocrine function. *In* Frontiers in Neuroendocrinology L. Martini and W. Ganong, eds. Pp. 237–290. New York: Oxford University Press.

Gottschalk, L., S. Kaplan, G. Gleser, and C. Winget
    1962 Variations in magnitude of emotion: A method applied to anxiety and hostility during phases of the menstrual cycle. Psychosomat. Med. 24:300–311.

Grant, E. C., and E. Mears
    1967 Mental effects of oral contraceptives. Lancet 2:945–946.

Green, R., W. Luttge, and R. Whalen
    1969 Uptake and retention of tritiated estradiol in brain and peripheral tissues of male, female, and neonatally androgenized female rats. Endocrinology 85:373–378.

Greene, R. and K. Dalton
    1953 The premenstrual syndrome. Brit. Med. J. 1:1007–1013.

Grounds, A. D.
    1965 Factors affecting symptoms arising during use of a new progestin in general practice. *In* Recent Advances in Ovarian and Synthetic Steroids, R. P. Shearman, ed. Pp. 185–197. Chicago: G. D. Searle.

Guthrie, R. D.
    1970 Evolution of human threat display organs. *In* Evolutionary Biology, T. Dobzhansky, M. K. Hecht, and W. Steere, eds. New York: Appleton, Century, Crofts.

Halasz, B.
    1969 The endocrine effects of isolation of the hypothalamus from the rest of the brain. *In* Frontiers in Neuroendocrinology, L. Martini and W. Ganong, eds. New York: Oxford University Press.

Hall, K. R., and I. DeVore
    1965 Baboon social behavior. *In* Primate Behavior, I. DeVore, ed. Pp. 53–110. New York: Holt, Rinehart and Winston.

Hamburg, D. A.
    1963 Emotions in the perspective of human evolution. *In* Expression of Emotion in Man. P. H. Knapp, ed. Pp. 300–317. New York: International University Press.

Hamburg, D. A., and D. T. Lunde
    1966 Sex hormones in the development of sex differences in human behavior. *In* The Development of Sex Differences. E. Maccoby, ed. Stanford, Ca.: Stanford University Press.

Hamburg, D. A., R. H. Moos, and I. D. Yalom
    1968 Studies of distress in the menstrual cycle and the post-partum period. *In* Endocrinology and Human Behavior, R. P. Michael, ed. Pp. 94–116. London: Oxford University Press.

Hansman, C., and M. Maresh
    1961 A longitudinal study of skeletal maturation. Amer. J. of Diseases of Children 101:305–321.

Harlow, H.
    1965 Sexual behavior in the rhesus monkey. *In* Sex and Behavior, F. Beach, ed. Pp. 234–265. New York: Wiley.

Harris, G. W., and D. Jacobsohn
    1951–1952 Functional grafts of the anterior pituitary gland. Proc. Roy. Soc. London, 139:263–276.

Haynes, H., B. L. White, and R. Held
1965 Visual accommodation in human infants. Science 148:528–530.

Herrmann, J. B., and F. E. Adair
1946 The effects of testosterone propionate on carcinoma of the female breast with soft tissue metastasis. J. Clin. Endocrin. Metab. 6:769–775.

Hinde, R. A.
1974 Biological Bases of Human Social Behaviour. New York: McGraw-Hill.

Hutt, C.
1972 Male and Female. Middlesex, England: Penguin Books.

Hutt, S. J., C. Hutt, H. G. Lenard, H. Bernuth, and W. J. Mantjewerff
1968 Auditory responsivity in the human neonate. Nature 218:888–890.

Iliff, A., and V. A. Lee
1952 Pulse rate, respiratory rate, and body temperature of children between two months and eighteen years of age. Child Development 23:237–245.

Ismail, A. A., R. A. Harkness, and J. A. Loraine
1968 Some observations on the urinary excretion of testosterone during normal menstrual cycle. Acta Endocrinol (Kobenhaun), cit. no. 3231410, 55:685–695.

Jacklin, C., and E. Maccoby
1972 Sex difference in intellectual abilities: a reassessment and a look at some new explanations. Presented at AERA, April, 1972.

Jacobs, T. J., and E. Charles
1970 Correlation of psychiatric symptomatology and the menstrual cycle in an outpatient population. Amer. J. Psychiat. 126:1504.

Janiger, O., R. Riffenburgh, and R. Kersh
1972 Cross-cultural study of premenstrual symptoms. Psychosomatics 13:226–235.

Janowsky, D. S., R. Gorney, P. Castelnuevo-Tedesco, and C. B. Stone
1969 Premenstrual-menstrual increases in psychiatric hospital admission rates. Am. J. Obstet. Gynecol. 103:189–191.

Jensen, G. D.
1968 Sex differences in the development of independence of infant monkeys. Behaviour 30:1–14.

Jost, A.
1953 Problems of fetal endocrinology: the gonadal and hypophysical hormones. Recent Progress in Hormone Research 8:379–418.

1958 Embryonic sexual differentiation. In Hermaphroditism, General Anomalies and Related Endocrine Disorders, H. W. Jones and W. W. Scott, eds. Baltimore: Williams and Wilkins Co.

1972 A new look at the mechanisms controlling sex differentiation in mammals. Johns Hopkins Medical J. 130:38–53.

Kagan, J.
1971 Change and Continuity in Infancy. New York: Wiley.

Kagan, J., and M. Lewis
1965 Studies of attention in the human infant. Merrill-Palmer Quart. 11:95–122.

Kane, F., R. Daly, J. Ewing, and M. Keeler
1967 Mood and behaviour changes with progestational agents. Brit. J. Psychiat. 113:265–268.

Keller, S.
1972 Sex Roles. Presentation made at conference on Research Issues in the Biological and Cultural Bases of Sex Roles, Palo Alto, Ca. June, 1972.

Kessler, S. and R. Moos
1972 Phenotypic characteristics of the XYY male. Comments on Contemp. Psychiat. 1:105–112.

1973 Behavioral manifestations of chromosomal abnormalities. Hospital Practice 8:131–137.

Kimura, D.
1967 Functional assymmetry of the brain in dichotic listening. Cortex 3:163–178.

1969 Spatial localization in left and right visual fields. Canad. J. Psychol. 23:445–458.

Knox, C., and D. Kimura
1970 Cerebral processing of non-verbal sounds in boys and girls. Neuropsychologia 8:227–237.

Kolodny, R. C., W. H. Masters, J. Hendryx, et al.
1971 Plasma testosterone and semen analysis in male homosexuals. N.E.J.M. 285:1170–1174.

Komarovsky, M.
1973 Cultural contradictions and sex roles: the masculine case. *In* Changing Woman in a Changing Society. J. Huber, ed. Pp. 111–122. Chicago: University of Chicago Press.

Kopell, B. S., D. T. Lunde, R. B. Clayton, and R. H. Moos
1969 Variations in some measures of arousal during the menstrual cycle. J. Nerv. Ment. Dis. 148:180–187.

Korner, A.
1973 Sex differences in newborns with special reference to differences in the organization of oral behavior. J. Child Psychol. Psychiat. 14:19–30.

Kramp, J. L.
1968 Studies on the premenstrual syndrome in relation to psychiatry, Acta Psychiat. Scanda. Suppl., 203:261–267.

Kreuz, L. E., and R. M. Rose
1972 Assessment of aggressive behavior and plasma testosterone in a young criminal population. Psychosom. Med. 34:321–332.

Kreuz, L. E., R. M. Rose, and J. R. Jennings
1972 Suppression of plasma testosterone levels and psychological stress: a longitudinal study of young men in officer candidate school. Arch. Gen. Psychiat. 26:479–482.

Kummer, H.
1968 Social Organization of the Hamadryas Baboon. Chicago: University of Chicago Press.

205

Kutner, S. J., and W. L. Brown
1972 History of depression as a risk factor for depression with oral contraceptives and discontinuance. J. Nerv. Ment. Dis. 155:163–169.

Laschet, U., and L. Laschet
1969 Three years clinical results with cyproterone acetate in the inhibiting regulation of male sexuality. Acta Endocinologica, Supp. 138:103.

Le Magnen, J.
1952 Les Pheromones olfacto-sexuel chez l'homme. Arch. Sci. Physiol. 6:125–160.

LeVine, R.
1970 Sex roles and economic change in Africa. In Black Africa: Its Peoples and Their Cultures Today. J. Middleton, ed. Pp. 174–180. London: MacMillan.

Levy, J., and T. Nagylaki
1972 A model for the genetics of handedness. Genetics 72:117–128.

Lewis, M., and S. Goldberg
1969 Perceptual-cognitive development in infancy: a generalized expectancy model as a function of mother-infant interaction. Merrill-Palmer Quart. 15:81–100.

Lewis, M., and M. Weintraub
1974 Sex of parent X sex of child: socio-emotional development. In Sex Differences in Behavior. R. Friedman, R. Richart, and R. Vande Wiele, eds. New York: Wiley.

Lisk, R. D.
1967 Sexual behavior: hormonal control. In Frontiers in Neuroendocrinology. L. Martini and W. Ganong, eds. New York: Academic Press.

Livingstone, F. B.
1967 The effect of warfare on the biology of the human species. In War: The Anthropology of Armed Conflict and Aggression, M. Fried, M. Harris, and R. Murphy, eds. Pp. 3–15. Garden City, N.Y.: Natural History Press.

Logothetis, J., R. Harnes, F. Morrell, and F. Torres
1959 The role of estrogens in catamenial exacerbation of epilepsy. Neurology 9:352–360.

Loraine, J. A., D. A. Adamopoulos, K. E. Kirkham, et al.
1971 Patterns of hormone excretion in male and female homosexuals. Nature 234:552–555.

Luttge, W. G.
1971 The role of gonadal hormones in the sexual behavior of the Rhesus monkey: a literature survey. Arch. Sex. Behav. 1:61–88.

Maccoby, E. E., and C. Jacklin
1974 The Psychology of Sex Differences. Stanford: Ca.: Stanford University Press.

Maccoby, E. E., and J. C. Masters
1970 Attachment and dependency. In Carmichael's Manual of Child Psychology (Third ed.), Vol. 2, P. Mussen, ed. New York: Wiley.

Maccoby, E. E., and W. C. Wilson
1957 Identification and observational learning from films. J. Abnorm. and Soc. Psychol. 55:76–87.

Mall, M. G.
1958 On the hormonal treatment of pre- and post-menstrual ovarian psychoses. *In* Psychoendocrinology, M. Reiss, ed. New York: Grune & Stratton.

Mandell, A., and M. Mandell
1967 Suicide and the menstrual cycle. J. Am. Med. Assoc. 200:792–793.

Martinez, C., and J. Bittner
1956 A non-hypophysical sex difference in estrous behavior of mice bearing pituitary grafts. Proc. Soc. Exper. Biol. and Med. 91:506–509.

McCammon, R. W.
1962 Some aspects of physical and physiological individual variation. Nutrit. Soc. Proc. 21:144–156.

McCarthy, D.
1954 Language development in children. *In* Manual of Child Psychology Second ed., L. Carmichael, ed. Pp. 492–630. New York: Wiley.

McClintock, M. K.
1971 Menstrual synchrony and suppression. Nature 229:244–245.

McGlone, J., and W. Davidson
1973 The relation between cerebral speech laterality and spatial ability with special reference to sex and hand preference. Neuropsychologia 11:105–113.

McGrew, W. C.
in   Evolutionary implications of chimpanzee sex differences, insect eating, and
press tool use. *In* Behavior of Great Apes. D. A. Hamburg and E. R. McCown, eds. Menlo Park, Calif.: Benjamin Cummings.

Meyer-Bahlburg, H., Dr. Rer Nat, D. A. Boon, M. Sharma, and J. A. Edwards
1974 Aggressiveness and testosterone measures in man. Psychosom. Med. 36: 269–274.

Michael, R. P.
1968 Gonadal hormones and control of primate behavior. *In* Endocrinology and Human Behavior. R. P. Michael, ed. Pp. 69–93. London: Oxford University Press.

1971 Neuroendocrine factors regulating primate behavior. *In* Frontiers in Neuroendocrinology. L. Martini and W. Ganong, eds. Pp. 359–398. London: Oxford University Press.

Michael, R. P., and E. B. Keverne
1968 Pheromones in the communication of sexual status in primates. Nature 218:746–749.

Michael, R. P., R. W. Bonsall, and P. Warner
1974 Human vaginal secretions: volatile fatty acid content. Science 186:1217–1219.

Michael, R. P., and J. M. Plant
1969 Contraception steroids and sexual activity. Nature 222:579–581.

Milner, B.
1974 Hemispheric specialization: scope and limits. *In* The Neurosciences: Third Study Program, F. Schmitt and F. Worden, eds. Cambridge, Mass.: MIT Press.

Mitchell, G. D.
  1969 Paternalistic behavior in primates. Psychol. Bull. 71:399–417.

Money, J.
  1970 Use of an androgen-depleting hormone in the treatment of male sex offend-
        ers. J. of Sex Research 6:165–172.

  1972 Developmental differentiation. *In* Man and Woman, Boy and Girl. J. Money
        and A. A. Ehrhardt, eds., Baltimore: Johns Hopkins University Press.

Money, J., and A. A. Ehrhardt
  1968 Prenatal hormone exposure: possible effects on behavior in man. *In* Endo-
        crinology and Human Behavior. R. P. Michael, ed. London: Oxford Uni-
        versity Press.

  1971 Gender dimorphic behavior and fetal sex hormones. (Presentation at Lauren-
        tian Hormone Conference, Sept., 1971).

  1972 Man and Woman, Boy and Girl. Baltimore: Johns Hopkins University
        Press.

Money, J., and J. Pollitt
  1964 Cytogenetic and psychosexual ambiguity. Arch. Gen. Psychiat. 11:589–
        595.

Moore, T.
  1967 Language and intelligence: a longitudinal study of the first eight years.
        Human Dev. 10:88–106.

Moos, R., B. Kopell, F. Melges, I. Yalom, D. Lunde, R. Clayton, and D. Hamburg
  1969 Fluctuations in symptoms and moods during the menstrual cycle. J.
        Psychosom. Res. 13:37–44.

Moss, H. A.
  1967 Sex, age and state as determinants of mother-infant interaction. Merrill-
        Palmer Quart. 13:19–36.

Moyer, K. E.
  1974 Sex differences in aggression. *In* Sex Differences in Behavior. R. Friedman,
        R. Richart, and R. Vande Wiele, eds. New York: Wiley.

Naftolin, F., K. J. Ryan, and Z. Petro
  1971 Aromatization of androstenedione by the diencephalon. J. Clin. Endocrin.
        Metab. 33:368–370.

Nebes, R. D.
  1974 Hemispheric specialization in commissurotomized man. Psychol. Bull.
        81:1–14.

Neumann, F., H. Steinbeck, and J. D. Hahn
  1970 Hormones and brain differentiation. *In* The Hypothalamus. L. Martini, M.
        Motta, and F. Fraschini, eds. Pp. 569–603. New York: Academic Press.

Newton, N.
  1973 Interrelationships between sexual responsiveness, birth and breast feeding.
        In Contemporary Sexual Behavior: Critical Issues for the 1970s. J. Zubin and
        J. Money, eds. Baltimore: Johns Hopkins University Press.

Oetzel, R. M.
   1966 Classified summary of research in sex differences. *In* The Development of Sex Differences. E. E. Maccoby, ed. Stanford, Ca.: Stanford University Press.

Oppenheimer, V. K.
   1973 Demographic influence on female employment and the status of women. *In* Changing Women in a Changing Society. J. Huber, ed. Pp. 184–199. Chicago: University of Chicago Press.

Paige, K.
   1971 Effects of oral contraceptives on affective fluctuations associated with the menstrual cycle. Psychosom. Med. 33:515–537.

Parlee, M. B.
   1973 The premenstrual syndrome. Psychol. Bull. 80:454–465.

Patterson, G. R., R. A. Littman, and W. Bricker
   1967 Assertive behavior in children: a step toward a theory of aggression. Mongr. Soc. Res. Child Devel. 32:5 and 6.

Persky, H.
   1974 Reproductive hormones, moods and the menstrual cycle. *In* Sex Differences in Behavior. R. Friedman, R. Richart, and R. Vande Wiele, eds. Pp. 455–466. New York: Wiley.

Persky, H., K. D. Smith, and G. K. Basu
   1971 Relation of psychologic measures of aggression and hostility to testosterone production in man. Psychosom. Med. 33:265–277.

Pfeiffer, C. A.
   1936 Sexual differences of the hypophysis and their determination by the gonads. Amer. J. Anatomy 58:195–226.

Phoenix, C. H., R. W. Goy, and J. A. Resko
   1968 Psychosexual differentiation of androgenic stimulation. *In* Perspectives in Reproduction and Sexual Behavior, M. Diamond, ed. Pp. 33–49. Bloomington, Indiana: Indiana University Press.

Porteus, S. D.
   1965 Porteus Maze Test: Fifty Years' Application. Palo Alto, Ca.: Pacific Books.

Raisman, R.
   1974 Evidence for a sex difference in the neuropil of the rat preoptic area and its importance for the study of sexually dimorphic functions. *In* Aggression. S. H. Frazier, ed. Pp. 42–51. Baltimore: Williams and Wilkins.

Raisman, G., and P. M. Field
   1971 Sexual dimorphism in the preoptic area of the rat. Science 173:731–733.

   1973 Sexual dimorphism in the neuropil of the preoptic area of the rat and its dependence on neonatal androgen. Brain Research 54:1–29.

Ransom, T. W., and B. S. Ransom
   1971 Adult male-infant relations among baboons (Papio anubis). Folia Primatol. 16:179–195.

Reddy, V. V., F. Naftolin, and K. J. Ryan
   1973 Aromatization in the central nervous system of rabbits: effects of castration and hormone treatment. Endocrinol. 92:589–594.

Ribeiro, A. L.
1962 Menstruation and crime. Brit. Med. J. 1:640.

Rivarola, M. A., J. M. Saez, W. J. Meyer, et al.
1966 Metabolic clearance rate and blood production rate of testosterone and androst 4-ene-3, 17-dione under basal conditions, ACTH and HCG stimulation: a comparison with urinary production rate of testosterone. J. Clin. Endocrinol. 26:371–378.

Rosaldo, M. Z.
1974 Women, culture, and society: a theoretical overview. *In* Women, Culture and Society, M. Z. Rosaldo and L. Lamphere, eds. Pp. 17–42. Stanford, Ca.: Stanford University Press.

Rosaldo, M. Z., and L. Lamphere
1974 Woman, Culture and Society. Stanford, Ca.: Stanford University Press.

Rose, R. M., I. S. Bernstein, and J. W. Holaday
1971 Plasma testosterone, dominance rank, and aggressive behavior in a group of male Rhesus monkeys. Nature 231:366–368.

Rosenblatt, J. S.
1965 Effects of experience on sexual behavior in male cats. *In* Sex and Behavior. F. Beach, ed. Pp. 416–439. New York: Wiley.

1967 Nonhormonal basis of maternal behavior in rats. Science 156:1512–1514.

Rosenblum, L.
1961 The development of social behavior in Rhesus monkeys. Ph.D dissertation, Psychology Department, University of Wisconsin.

1974 Sex differences in mother-infant attachment in monkeys. *In* Sex Differences in Behavior. R. Friedman, R. Richart, and R. Vande Wiele, eds. New York: Wiley.

Rossi, A.
1974 Physiological and social rhythms: the study of human cyclicity. Special lecture to American Psychiatric Association, Detroit, Michigan, May, 1974.

Ryan, K. J., F. Naftolin, V. Reddy, F. Flores, and Z. Petro
1972 Estrogen formation in the brain. Am. J. Obstet and Gynecol. 114:454–460.

Sanday, P. R.
1974 Female status in the public domain. *In* Women, Culture and Society. M. Z. Rosaldo and L. Lamphere, eds. Pp. 189–206. Stanford, Ca.: Stanford University Press.

Saunders, F. G.
1968 Effects of sex steroids and related compounds on pregnancy and on development of the young. Physiol. Rev. 48:601–643.

Schaffer, H. R., and P. E. Emerson
1964 The development of social attachments in infancy. Monogr. Soc. Res. Child Devel. 29:5–77.

Schwartz, N.
 1968 Newer concepts of gonadotrophic and steroid feedback control mechanisms.
   In Textbook of Gynecologic Endocrinology. J. J. Gold, ed. Pp. 33–50. New
   York: Hoeber.

Sears, R. R.
 1963 Dependency motivation. *In* The Nebraska Symposium on Motivation.
   M. R. Jones, ed. Pp. 25–64. Lincoln, Nebraska: University of Nebraska
   Press.

Segal, S. J., and D. Johnson
 1959 Inductive influence of steroid hormones on the neural system: ovulation
   controlling mechanisms. Archives d'Anatomie Microscopique et de Mor-
   phologie Experimentale 48:261–274.

Sherman, J. A.
 1971 On the Psychology of Women. Springfield, Ill.: Charles C. Thomas.

Simpson, G. G.
 1958 The study of evolution methods and present status of theory. *In* Behavior and
   Evolution. A. Roe and G. G. Simpson, eds. Pp. 7–26. New Haven: Yale
   University Press.

Sommer, B.
 1973 The effect of menstruation on cognitive and perceptual-motor behavior: a
   review. Psychosom. Med. 35:515–534.

Spencer, B., and F. J. Gillen
 1927 The Arunta: A Study of a Stone-Age People. London: Macmillan.

Sperry, R. W.
 1974 Lateral specialization in the surgically separated hemispheres. *In* The
   Neurosciences: Third Study Program. F. Schmitt and F. Worden, eds. Pp.
   5–19. Cambridge, Mass.: MIT Press.

Sperry, R. W., and J. Levy
 1970 Mental capacities of the disconnected minor hemisphere following commissurotomy. Presented at A.P.A., at the symposium "Asymmetrical Function
   of the Human Brain" Miami, Florida.

Steward, J. H.
 1936 The economic and social basis of primitive bands. *In* Essays in Anthropology
   Presented to A. L. Kroeber. R. H. Lowie, ed. Berkeley, Ca.: University
   of California Press.

Tanner, J. M.
 1960 Genetics of human growth. *In* Human Growth. J. M. Tanner, ed. New
   York: Pergammon Press.

 1970 Physical growth. *In* Carmichael's Manual of Child Psychology. Third ed., P.
   Mussen, ed. New York: Wiley.

Taylor, D. C.
 1969 Differential rates of cerebral maturation between sexes and between hemispheres. The Lancet July 19, pp. 140–142.

Taylor, S. P., and S. Epstein
 1967 Aggression as a function of the interaction of the sex of the aggressor and the
   sex of the victim. J. of Personality 35:474–486.

Turiaf, J., P. Blanchon, and L. Zizine
 1949 Asthme et troubles menstruels menopause exlue. Bull. et Men. de la Societe
   Medicale des Hospitaux de Paris 65:347–353.

Tyler, L.
1963 Tests and Measurements. New Jersey: Prentice-Hall.

Udry, J. R., and N. M. Morris
1968 Distribution of coitus in the menstrual cycle. Nature 220:593–596.

U.S. Bureau of the Census, Pocket Data Book, U.S.A.
1971 Washington D.C.: U.S. Government Printing Office.

Valenstein, E. S., and W. C. Young
1955 An experiential factor influencing the effectiveness of testosterone proprion-
ate in eliciting sexual behavior in male guinea pigs. Endocrinol. 56:173–
177.

Van Lawick-Goodall, J.
1968 The behavior of free-living chimpanzees in the Gombe Stream area. *In* Animal Behavior Monographs. J. M. Cullen and C. J. Beer, eds. London: Bail-
liere, Tindall, and Cassell.

Vierling, J. S., and J. Rock
1967 Variations in olfactory sensitivity to exaltolide during the menstrual cycle. J.
Appl. Physiol. 22:311–315.

Ward, I.
1974 Sexual behavior differentiation: prenatal hormonal and environmental control. *In* Sex Differences in Behavior. R. Friedman, R. Richart, and R. Vande
Wiele, eds. Pp. 3–17. New York: Wiley.

Washburn, S. L.
1969 The evolution of behavior. *In* The Uniqueness of Man. J. Rolansky, ed.
Amsterdam and London: North Holland.

Washburn, S. L., and C. S. Lancaster
1968 The evolution of hunting. *In* Man the Hunter. R. B. Lee and I. DeVore,
eds. Pp. 293–303. Chicago: Aldine.

Washburn, S. L., and E. R. McCown
1972 Evolution of human behavior. Social Biol. 19:163–170.

Washburn, S. L., and E. R. McCown
1974 The new science of human evolution. *In* 1974 Brittanica Yearbook of Science
and the Future. Pp. 33–49. Chicago: Encyclopedia Brittanica, Inc.

Waxenberg, S. E.
1963 Some biological correlates of sexual behavior. *In* Determinants of Human
Sexual Behavior. G. Winokur, ed. Pp. 52–75. Springfield, Ill.: Charles C.
Thomas.

Wetzel, R. D., T. Reich, and J. N. McClure
1971 Phase of the menstrual cycle and self-referrals to a suicide prevention service.
Brit. J. Psychiat. 119:525–526.

Whalen, R.
1968 Differentiation of neural mechanisms which control gonadotropin secretion
and sexual behavior. *In* Perspectives in Reproduction and Sexual Behavior.
M. Diamond, ed. Bloomington, Ind.: Indiana University Press.

1974 Sexual differentiation: models, methods, and mechanisms. *In* Sex Differences in Behavior, R. Whalen, R. Richart, and R. Vande Wiele, eds. Pp.
467–481. New York: Wiley.

White House Conference on Youth.
1970 Washington, D.C.: U.S. Government Printing Office.

Witelson, S. F., and W. Pallie
1973 Left hemispheric specialization for language in the newborn. Brain 96:641–646.

Witkin, H. A., R. B. Dyk, H. F. Faterson, D. R. Goodenough, and S. A. Karp
1962 Psychological Differentiation. New York: Wiley.

Witkin, H. A., H. B. Lewis, M. Herzman, K. Machover, P. B. Meissner, and S. Wapner
1954 Personality through Perception. New York: Harper and Row.

Wolstenholme, G., and C. O'Conner
1962 The Life Span of Animals, Ciba Foundation Colloquia on Aging, Vol. 5, Boston: Little, Brown.

Yalom, I., R. Green, and N. Fisk
1973 Prenatal exposure to female hormones: effect on psychosexual development in boys. Arch. Gen. Psychiat. 28:554–561.

Yarrow, L. J., and F. A. Pedersen
1972 Attachment: its origins and course. In The Young Child: Reviews of Research, Vol. 2. W. W. Hartup, ed. Pp. 54–66. Washington, D.C.: National Association for the Education of Young Children.

R. V. S. Wright

Imitative Learning of a Flaked Stone Technology—The Case of an Orangutan

There is a long history of speculation about the nature of the earliest toolmakers. Could a small-brained hominid have made tools? Did both forms of Australopithecus make tools? Must the hand have evolved in a human direction before tools could be made? Wright is the first to have seen that experiments might elucidate the nature of the problem. In this chapter he describes how an orangutan is influenced to make and use simple stone tools. Wright shows that experiments may be critical in the understanding of the problems of human evolution.

INTRODUCTION

Let us suppose that Jane Goodall had seen a group of apes striking flakes from a core and using these flakes as cutting tools. How would the news have been received by anthropologists? Speculating on this question led me to devise the experiment described on the following pages.

This is a revised and slightly expanded version of a paper of the same title published in *Mankind* VIII (1972), pp. 296–306. I am indebted to the Anthropological Society of New South Wales for permission to reproduce it here.

The Behavioral Significance of Stone Toolmaking

Flaked stone tools are made from siliceous rocks that have remarkable survival properties of physical and chemical stability. Flaking technology went hand in hand with much wastage and throwing away for replacement. Thus, all over the world archaeologists find a great litter surviving from the prehistoric past. Flaked stone artefacts are studied as cultural traces not only because they are plentiful, but also because there are, at times, literally no other things surviving. We choose to wring the maximum of information out of these relics.

A stone litter is often the only sign of activity surviving to the present from those remote times of the early hominids. Anthropologists have evaluated its behavioral significance in various ways. Some have wanted to include the ability to make stone tools in the characters to be used in the zoological taxonomy of the Hominidae (Wells 1959). A less formal view, and a more common one, is to see the ability as in some way a special *human* quality. Holloway reflects this attitude when attributing stone tools to australopithecines:

> the stone tools show that finer degrees of manual dexterity had been attained by the time of the australopithecines. The stone tools also indicate that these hominids were capable of organizing experience in a more human manner. [1968:185]

The apes are unwelcome applicants for admission to the human club and they can be effectively blackballed by their inability to flake stone. Therefore, the sharpest reaction to my imaginary discovery would be from those who need a single and simple qualitative trait to be stamped, as a hallmark, onto the complex notion of *human*. If apes had been seen flaking stone in the wilds, we can imagine the rules of entry being subsequently altered to require apes to pass a revised qualitative test—perhaps the ability to execute trimming on the struck flakes.

Jane Goodall's fascinating descriptions of termiting among chimpanzees did, in fact, rock the classificatory boat. Her apes were making twigs into tools, and the ability to make tools had been set as a human hallmark. The rocking is lessened by calling the prepared twigs "very crude tools" (Goodall 1965:473)—which indeed they are. But would we categorize the stone flakes and cores of our imaginary apes in the same way? I think not. This would be because we rank stone flaking as requiring far greater understanding and manipulative skill than termiting. Indeed, some have thought that stone flaking might have been too difficult even for the australopithecines of Africa—hominids, assumed, by common consent, to have been more intelligent than any of the living apes.

Inskeep distills these attitudes about one species of *Australopithecus* when he says:

> Few people would demur at the proposal that *A. africanus* was a tool-user, and it would appear that there are equally few who would regard him as a tool-maker in the sense in which the habit is associated with man. [1969:177]

Thus, where australopithecine fossils have been found with tools of flaked stone, there has at times been some disquiet, followed by a sense of relief, when the responsibility could be lifted from australopithecine shoulders. Discussing the South African sites of Swartkrans and Sterkfontein Mary Leakey has said:

> The question of whether *Homo erectus* or *Australopithecus* was responsible for these tools has been discussed at length by various authors. . . . Arguments in favour of one or other taxon (or both) being toolmakers continued to be largely hypothetical until 1969 when several cranial fragments from Swartkrans . . . were fitted together by R. J. Clark. . . . The assembled pieces leave little doubt that the specimen belonged to the genus *homo* although not necessarily *Homo erectus*. [1971:280−281]

Australopithecines as Stone Toolmakers

At the Olduvai Gorge the Leakey family has recovered data that is of outstanding interest in the study of human evolution. Out of this wealth of data and interpretation I shall extract what is really an ethological problem, on which the experiment I shall describe casts some indirect light.

One of the interpretations of the fossil evidence at the Olduvai Gorge is that there coexisted, around the shores of the ancient Olduvai Lake, two species of hominids. They coexisted for a period of about one million years. One of these species was a robust australopithecine (for example, *Zinjanthropus*) and the other a more gracile hominid named *Homo habilis*. The australopithecines are thought to have been less intelligent than *Homo habilis*. These ideas have been developed over the last decade's work at the Olduvai Gorge.

It is less widely known that the idea of distinct species of hominids, sympatric, and differing in intelligence, goes back several years to when Broom and Robinson reported on the cave of Swartkrans in South Africa (1950). Both in eastern and southern Africa, flaked stone artefacts have been found in the same deposits as these hominids. Statements on who made them have been hedged with cautionary qualifications. Mary Leakey says, "Whether *Zinjanthropus* or the co-existent *Homo habilis* was responsible for making the tools and occupying the floor remains an open question although the balance of the evidence favours Homo habilis" (1971:49). See also Leakey and Goodall (1969:149) for a summary of the alternative responsibilities.

Mary Leakey has elsewhere expressed herself rather more strongly against australopithecine capability; "it seems unlikely that he was responsible for more than tool-using, or possibly for simple modification of objects without employing another instrument for the purpose." (1971:272). P. Tobias (1971) has brought up to date his earlier useful surveys of this vexatious question.

A Proposition

Let us take for examination the hard-line view (with qualifications removed) that there coexisted in Africa for some hundreds of thousands of years two species of hominids and that stone tools were made by one species but not by the other. This restriction of stone flaking to *Homo habilis* could have a number of explanations. For instance:

1. the flaking of stone was species-specific behaviour and unlearned, like nest-building in birds. I find this improbable in the light of studies in comparative primate behaviour.
2. More plausibly, the niches of the two species were different and only *H. habilis* had a system that incorporated stone tools.
3. Australopithecines did not have the brains and manipulative skills to make stone tools—they were unable to master the procedure by either self-discovery or by imitative learning from more advanced hominids.

My experiment relates to the *imitative learning* part of the third explanation. How the two species of hominids normally interacted on a day-to-day basis scarcely matters in the behavioral model proposed. This is because of the great duration of their coexistence. For my purpose I can think of their interaction in the language of astronomical probability. Let us assume that it was an exceedingly rare event for an australopithecine to actually see a *Homo habilis* flaking stone—perhaps, for the sake of argument, it happened only once a year in the whole of Africa. There would, even so, have been a million such sightings during the period of coexistence that has been defined archaeologically.

Propositions on australopithecine capability cannot, of course, be subjected to direct behavioral tests. They can, however, be examined in two ways. One of these is to accumulate more associations between the hominid fossils and artefacts—and this is in the indefatigable hands of the Leakeys and their colleagues.

Another way suggests itself—extrapolation from the imitative be-haviour of living apes. The imitative capabilities of chimpanzees are exploited by circuses and space scientists. As far as I know, apes have not been allowed to show their paces at making stone implements. My experiment makes a start in this direction.

THE EXPERIMENT

Aim

The aim of the experiment was to get an ape, by imitative learning, to use a hammerstone to strike a flake from a flint core, and to use this flake as a tool to open a box.

The Programme of the Experiment

The programme was divided into two stages. In each stage the ape was given demonstrations of what to do and was then given the opportunity of doing it himself. By imitative learning the ape was to become a tool user in Stage I and a toolmaker in Stage II.

The aim of Stage I was to get an ape to cut a cord to open a box. He was to use a flint flake—one made by me out of his sight. There is nothing special about such a flake. It can be used for generalized cutting in the same way that we use knives. For a striking fictional analogy of Stage I see E. A. Poe's *The Murders in the Rue Morgue*.

The aim of Stage II was to get an ape to make his own flakes and use them to open the box.

The programme was thus explicitly a test of imitative learning and not of intuitive problem solving. More pretentiously, the experiment was a case of interspecies cultural transmission of a lithic technique within the family Hominoidea.

The Animals

I thought first of using chimpanzees or gorillas because I assumed that they were more likely to succeed than gibbons or orangutans. However, matters of housing and pregnancy at the Bristol Zoo forced me to use orangutans. I was prejudiced against members of this species on two grounds—their reportedly arboreal habits (Schaller 1965) and the extreme shortness of their thumbs. Had I been aware of the behavioural implications of the wide Asiatic spread of the genus in the Pleistocene, I would not have approached it with such pessimism in the Bristol Zoo!

There was a brief period when I used two juvenile orangutans in the cage together, but, because they distracted each other, I soon used only one of them, Abang (Figure 1). He was about five and a half years old. When about a year old he was illegally captured in Sarawak. Having been confiscated by

FIGURE 1
Abang ready to start flaking.

the authorities, he was taken into official care and flown with a female, Dayang, to England in 1967 as a donation from the Orangutan Rescue Scheme. During their stay in Bristol Zoo neither of these orangutans had been used in experiments to test manipulative skills.

Equipment

The box is made of sheet aluminum so that the ape could not damage it by violence. On top there is a hinged lid and set in this lid is a window of wire mesh through which the ape could see food. In the side of the box, at one end, is a slot; the animal could put its hand into the box through this slot (Figure 2) but it could not reach the food that was behind a vertical partition. The only way to get the food is to open the lid but this is secured by a nylon cord fastened *inside* the box between the lid and the floor. Therefore, the ape had to put its hand into the slot and cut through the cord before it could open the lid.

FIGURE 2
Preparing to cut the cord inside the slot. The loose end of cord hangs down from the clamp on top of the box.

221

The flint collected was an assortment of nodules from a chalk quarry near Westbury (Wiltshire). This same flint was widely used in prehistoric times for the manufacture of implements.

The hammerstone was a river pebble of quartzite weighing 1.4 kg (Figure 1).

The Experimental Environment

The experiments were carried out in the indoor part of the ape's cage. I was working in a busy zoo with a valuable animal. Naturally I had to accept certain restrictions and arrangements that were unsatisfactory by the standards of a behavioral laboratory. For example, the orangutan was sometimes distracted by people outside his cage and the wooden floor of the cage was too unyielding a surface on which to steady an irregular core of flint.

It is impossible to write notes when in the same cage as an ape so I used a cassette recorder concealed under my jacket.

Explanation of Timetable

The two stages were carried out in a number of sessions and for ease of reference I have divided the sessions into episodes. In the table below is a skeleton of the events. These are filled out in the subsequent *Comments*.

Episodes have an identification number, for example 3/5 signifies the 3rd session/5th episode. *Demonstration* means, in Stage I, that I used a flake to cut the cord to open the box; in Stage II it means that I struck a flake from a core and that this was used by myself or the ape to cut the cord to open the box.

During an *exposure* the ape had the opportunity of imitating a demonstration. Exposures are classed as *failures* if the ape had not successfully imitated the demonstrations when the episode was terminated. Exposures are classed as successful if the ape successfully imitated the demonstration.

Comments on Stage I

1st Session. Abang and Dayang were in the cage for the whole session. In between my demonstrations they put the flint to the slot five times, Abang on two of these occasions making a sawing motion. There is no reason to think that he was specifically imitating me; his keeper says that the apes

TABLE 1
Stage 1 of the Experiment

SESSION NO. AND DATE	EPISODE NO.	APPROX. LENGTH OF EPISODE (MINUTES)	NATURE OF EPISODE	OUTCOME	NOTES
1st Session	1/1	15	Exposure	Failure	Box fails
5 March 1971	1/2	3	Demonstration	—	
	1/3	2	Demonstration	—	
	1/4	2	Exposure	Failure	Box fails
	1/5	8	Exposure	Failure	
	1/6	3	Demonstration	—	
	1/7	4	Exposure	Failure	Box fails
	1/8	2	Exposure	Failure	
	1/9	< 1	Demonstration	—	
		40			
2nd Session	2/1	1	Exposure	Failure	
11 March	2/2	3	Demonstration	—	
	2/3	4	Exposure	Failure	Box fails
	2/4	2	Exposure	Failure	
	2/5	3	Demonstration	—	
	2/6	2	Exposure	Failure	
	2/7	2	Demonstration	—	
	2/8	3	Exposure	Failure	
	2/9	1	Exposure	Failure	
	2/10	2	Demonstration	—	
	2/11	4	Exposure	Failure	
	2/12	2	Demonstration	—	
	2/13	4	Exposure	Success	
	2/14	< 1	Exposure	Success	
		34			
3rd Session	3/1	3	Exposure	Success	
16 March	3/2	3	Exposure	Success	
	3/3	5	Exposure	Success	
	3/4	4	Exposure	Success	
		15			
4th Session	4/1	5	Exposure	Success	
19 March	4/2	4	Exposure	Success	
		9			

commonly make such a reciprocating motion with a twig against the bars of the cage.

They got the fruit three times by *illicit* means when the box failed. They also disturbed each other in a way that lowered the chance of success. For example, in 1/5 Dayang swung the box round just as Abang was putting the flint into the slot.

TABLE II
Stage II of the Experiment

SESSION NO. AND DATE	EPISODE NO.	APPROX. LENGTH OF EPISODE (MINUTES)	NATURE OF EPISODE	OUTCOME	NOTES
5th Session	5/1	3	Exposure	Failure	
23 March	5/2	2	Demonstration	—	
	5/3	1	Exposure	Failure	Hammerstone breaks
	5/4	7	Demonstration	—	
	5/5	9	Exposure	Failure	
		22			
6th Session	6/1	2	Exposure	Failure	
25 March	6/2	2	Demonstration	—	
	6/3	9	Exposure	Failure	Core breaks loose
	6/4	1	Demonstration	—	
		14			
7th Session	7/1	1	Exposure	Failure	
29 March	7/2	1	Demonstration	—	
	7/3	12	Exposure	Failure	2 cores broken
	7/4	5	Demonstration	—	
		19			
8th Session	8/1	19	Exposure	Failure	Strikes chip
2 April		19			
9th Session	9/1	12	Exposure	Failure	Strikes chip
5 April	9/2	18	Demonstration	—	
		30			
10th Session	10/1	17	Exposure	Failure	Strikes flake
6 April	10/2	11	Exposure	Failure	Strikes flake
	10/3	2	Exposure	Success	Strikes flakes and opens box
		30			
11th Session 16 April	An unregulated filming session of about 1½ hours in which on four occasions Abang struck flakes and used them to open the box.				

2nd Session. Abang and Dayang were together in the cage up to and including episode 2/7; thereafter Abang was alone. I chose Abang on the hunch that he had more potential than Dayang.

The box had been modified to prevent equipment failure; the modifications were successful—the failure in episode 2/3 was due to faulty assembly.

Up to 2/6 Abang still saw forcing the lid or string as the best way of opening the box, not surprisingly, given the equipment failures. Nevertheless, in 2/6 he twice held the flint to the slot in a way that suggested he saw it as one solution.

It was not until 2/11 that his actions gave him any chance of cutting through the cord. But he still did many other things with himself and the equipment. As an example of this varied behavior I gave a summary of his actions in the four minutes of episode 2/11:

> Abang goes to the slot end of the box and looks in; he picks up the flake and rubs it against the cord. He pulls at the cord with his fingers and rubs the flake against the metal edges of the slot and then, reaching right inside the box, bangs on the floor with the flake. He drops it and stares at me. I give it back to him and he scratches it around on the floor inside the box. He then tries to force up the lid of the box, gives up, and fiddles around with Vernon Reynolds' [a visiting scholar] shoe laces. He is given the flake again and takes it over to water trapped in a knothole in the cage floor; he dips the flake in the water, sniffs it, bites it, and bangs it on the floor. He picks up the flake in his toes, sticks his foot into the slot, and drops the flake inside the box. He moves round to the side of the box and tries to force up the lid, then goes back to the slot end. He is given the flake again and rubs it against the cord in a way which would have severed it had he used a sharp edge. Casually he tries to lift the lid and then intersperses sniffing at the flake with sawing at the outside of the box. He puts an empty hand into the box, momentarily walks away, but returns to sit on the box where he does nothing.

In episode 2/12 his keeper guided his hand to cut the string. In retrospect it would clearly have been more informative had we been less helpful to Abang. I had, however, expected a long slow road to his mastering the technique and had certainly not anticipated the transformation in his capability that began in episode 2/13.

The essentials of this transformation can be appreciated by comparing his economy of action in episode 2/14 with his dispersed actions in 2/11 outlined above. In episode 2/13 he goes to the front of the box and tries to force up the lid. Then he picks up the flake, saws against the cord and very soon severs it. His economical behavior in this episode is characteristic of his behavior in later episodes of this stage. He has now drastically reduced the range of things he does with himself and the equipment.

3rd Session. I was anxious to see if his capability had slumped in the intervening five days. It had not. In episode 3/1 Abang went straight over to the box, looked into the slot, picked up the flake, and cut the cord. He did not hesitate and he set to work with the same manipulative control he showed at the end of the previous session.

While I was reloading the box at the end of episode 3/1 Abang tried to grab at the fruit before the lid was shut. His keeper restrained him. At this Abang squeaked a protest. After a moment he began to scream with rage and forced his way back to the box. His keeper, feeling constrained to discipline him, took the screaming animal to a tiled platform in the far corner of the cage and talked quietly to him. I felt upset and went over, offering Abang the flake. He took it, licked it and though still noisy, he was helped off the platform by his keeper. Then he went straight over to the box, becoming silent as he went. With three sawing cuts he severed the cord, opened the box and took out the fruit. He did the job without fumbling or hesitation.

These happenings in episode 3/2 are the more dramatic peaks of a continuous series of interactions between myself, Abang, his keeper, and assorted distant spectators. They are a special class of uncontrolled variables that would complicate assessment in comparative studies of capability.

4th Session. To reinforce earlier learning, I used a thicker cord. In episode 4/1 Abang worked for nearly two minutes trying to cut a slack cord with a blunt edge; he then cut his finger with the flint. It was not a bad cut (just a little nick) but it continued to ooze blood. He continued working until he had completed the job in another minute, intermittently licking the blood away from his finger.

In episode 4/2 he lost interest and put the flake down. Walking away from Abang, I took the flake over to the far corner of the cage and produced two more from my pocket. Abang came over and took the three flakes from my hand, one by one. He then went back to the box and for a couple of minutes alternated between trying to bite the cord and trying to cut it. During cutting he exchanged one flake for another.

Comments on Stage II

5th Session. In episode 5/1 Abang played with the hammerstone and flint core before I had given him any demonstration. For the whole three minutes he showed no interest in the box though it was pointed out to him. He had not been fed since the day before and was presumably as hungry as previously. Curiosity about novel equipment overrode a desire to feed himself.

Immediately after the demonstration in episode 5/2 (when I used the hammerstone to strike a flake with which I opened the box) Abang picked up the hammerstone and banged the core. I suspect that he enjoyed the noise and had not linked the new equipment to the solution of opening the box. In fact, in episode 5/3 he used the hammerstone to bang on the wall of the cage. When he finally became interested in the box he tried the impossible task of using the hammerstone on the cord.

He still seemed to prefer making a noise when, in episode 5/4, he used bits of the shattered hammerstone to bang the core; this was even after I had offered him the flake made at the demonstration. However, at the end of this episode he did suddenly pick up this flake and go and cut the cord with it.

By episode 5/5 he seemed to have lost interest in making a noise. He had also lost interest in the box, perhaps because he had no flake to open it with. For nine trying minutes he gave himself up to random play, twice pushing me over.

It is worth noting that when I broke the hammerstone in episode 5/4 I probably made it impossible for him to detach a flake with the small piece left over, even if he had persistently imitated my actions.

6th Session. Before episode 6/1 Abang opened the box twice, with no hesitation, but using a ready-made flake. I had arranged this performance for Drs. Jonathan Musgrave and Vernon Reynolds. Abang's unprompted application and dexterity confirmed my feeling that when he had fooled around in the previous session it was not because he had forgotten that he could open the box with a flake.

Abang did hit at the core in episodes 6/1 and 6/3. However the spectrum of his actions was still so wide that I do not think that he had grasped the relationship between what he was doing and opening the box. For instance, toward the end of episode 6/3, after hitting the core, he, in sequence:

> levers at the lid with his incisors, lies on his back with the box upside down on his face, pushes the cord with the tip of a broken stick, lies on his back rolling the hammerstone between his feet and hands, balances the hammerstone on the back of his neck, puts the hammerstone in his motor tyre and rolls both around the cage, pushes the hammerstone against the cord, drops the hammerstone into the box, struggles to retrieve it, and uses the core to try to cut the string.

Once the core had been broken out of the plaster in episode 6/3 he was unlikely to have detached a flake by hitting the then unstable core.

7th Session. Failings in the equipment again lessened the chances of success. This time I tried cores stuck on wood with denture repair cement and with epoxy resin. Abang broke both mountings, leaving himself with unstable cores.

When the first core broke from its mount in episode 7/3 it left a splinter of flint lying by the box. Abang either did not see it or he ignored it. He tried other methods of opening the box and, for the first time, showed apparent feverish frustration. He finally managed to bite through an already frayed cord—a satisfying personal success if not an experimental one.

227

After the demonstration in 7/4 Abang made an innovation. The slot having been narrowed after 7/3 so that he could no longer bite the string, he took one of the flakes between his *teeth* and pushed it against the cord. But he soon stopped this and took up another flake in his hand with which he cut the cord.

I assume that he enjoyed doing new things with the equipment and that he did not mechanically go for the most rapid way of getting the fruit. It is arguable how, on any particular occasion, one can distinguish between choice on the animal's part and lapses in his understanding. After watching him through the sessions, I have a strong impression that unless some unaccountable switch is turning his capability on and off, he, at times, chooses not to fill his empty stomach even though he is faced with the box filled with fruit and the means of opening it.

8th Session. This time the cores were strapped onto a board with a piece of plastic foam between. They again proved unstable. Abang showed much more inclination to hammer in this session than in any previous one.

Early on he detached a splinter of flint which he ignored at the time. Ten minutes later, while rolling the hammerstone around the floor of the cage, he picked this splinter up in his lips, seemed to lodge it against his lower incisor teeth, and then pressed it against the cord for a few seconds; the splinter fell out of his mouth into the box. My recollection is that he dropped it after I had, for some spontaneous and unthinking reason, made a discouraging signal.

9th Session. The cores were set in expanded polyurethane foam but were again disappointingly unstable.

This was an inconclusive session during which I began to despair of the combination of Abang and the equipment. While hammering, he did detach a minute piece of soft cortex from the core and immediately tried to use it on the cord—an action which was foredoomed to failure.

My final demonstration in 9/2 may have made a special impression on him. I knocked off three flakes in as many seconds. Abang, of his own accord, immediately tried one on the cord but failed. He went across the cage to get another one that had shot over there. Armed now with a flake in each hand he settled down and cut the cord.

10th Session. Success came in this session. That it did not come earlier was undoubtedly due to my tardiness in solving the problem of unstable cores. The solution was to use a cushion of modelling clay (Plasticine) between the core and a board; after the core had been pressed into the warmed clay, it was secured with a leather strap around the lot. The advantage of this method (over the ingenuities of modern chemical adhesives) is that there is no rigid bond to be fractured. Furthermore the clay can, during the session itself, be pressed up again under the core if it becomes loose.

The box also had his whole daily meal in it instead of pieces of fruit. It is unlikely, however, that the increased food contributed to the final success because during the session he engagingly fooled around as usual.

In this session he struck four flakes in three episodes (Figure 3). In each episode he immediately, and without any encouragement or conscious sign from myself or his keeper, picked up the flake and went to use it on the cord. In episodes 10/1 and 10/2 he failed to cut the cord.

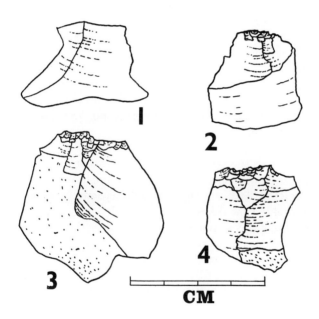

FIGURE 3
Dorsal surfaces of flakes struck by Abang during the tenth session. Striking platforms face upwards. Fine flaking at the platform edge is due to crushing from prior mis-hits at the edge of the core.

I give a precis of the events in episode 10/3—the last documented session:

> Having failed to cut the cord in episode 10/2 Abang puts the flake he struck in that episode into his mouth (Figure 3:2). Then he spits it out and goes over to the core on his own accord. During eight blows he knocks off two flakes. Without encouragement or conscious signal from myself or his keeper he immediately picks up the two flakes and takes them over to the box. Using the larger flake (Figure 3:3) in his left hand, he gives eleven vigorous cuts on the cord. Momentarily he puts the flake in his mouth and then takes it out again to give several additional cuts. He changes to his right hand for a few more. Some strands still remain to be severed and to do this he changes to the smaller flake (Figure 3:4), previously unused. The time from detachment of the flakes to severing of the cord is 80 seconds of continuous work. He takes his armful of rewards from the box and, while eating them, urinates over the core.

He developed a new practice in this session and pursued it in the next. Temporarily holding the flake between his lips, he took a fresh grip on it with his fingers, sometimes turning it to produce a new edge. He may have begun to appreciate different degrees of sharpness around a flake, but to test this hypothesis would require a number of observations on flakes with marked edges.

I formed the impression in this session and in the next that he had correctly realized that to detach flakes he had to hit around the edges of the core platform and not toward the center; again, this would need to be quantified to be conclusive.

11th Session. This session was not recorded except on film by the BBC. Abang fooled around for about ten minutes before he settled down to work. He then detached flakes on three or four separate occasions, each time immediately taking the flakes to cut the cord.

A brief film of this session survives, edited by the BBC. Among other shots it contains one uncut sequence showing detachment of a flake and its immediate use. Unfortunately, the unedited footage was not kept, owing to a misunderstanding at the BBC.

Further work was not possible because I had to return to Australia.

DISCUSSION

Manipulative Capability

Napier has outlined the functional duality of a young orangutan's hand (1960). The two grips which Napier describes were used by Abang. In the power grip he folded his hand around the hammerstone holding it in the palm and flexed fingers; his abducted thumb made some steadying contribution to this grip (Figure 4).

He cut with the apes' version of the precision grip, holding the flakes between the pulp of the thumb and the sides of the proximal and middle phalanges of the index finger (Figure 5). With his fingers tucked out of the way into his palm, he made a neat use of small flakes. However, he was in danger of cutting his index finger where skin folds bulged around the proximal end of the proximal phalanx. He did once cut himself there. The form of the human hand allows us to hold a flake out of such danger between the thumb and tips of the fingers.

The orangutan has a shorter thumb than a gorilla or chimpanzee. Its

FIGURE 4
Abang hammering at the core. He steadies the board with his foot and with the back of the flexed wrist of his left hand.

FIGURE 5
An ape's version of the precision grip.

hand is thus furthest from the human in this character. Yet it turned out to be a highly preadapted organ for the new function I set it. In fact, with its additional brachiating capabilities one could see an orangutan's hand as more generalized in its form and function than a man's.

It is a sobering exercise to think how easy it would be to devise an evolutionary argument that explained the orangutan's physical form in terms of the functions it was fulfilling in this experiment—for example, hammering away at a core steadied in its prehensile feet.

We have perhaps been misled by associating the arboreal state with low manipulative capability and the terrestrial state with a high one. Whatever the generality, Abang belies the idea that until you have been down from the trees for some time, you cannot have a sensitive tool-using and toolmaking capability.

Writing quite outside the ambit of this discussion, Carpenter and Durham (1969) have given us what I judge to be a correct lead in this problem. They give a preliminary report on a comparative study of the locomotory behavior of primates. Having discussed the problems of meaning in the word *brachiation,* they usefully divide locomotory behavior into that carried out supported on a substratum and that carried out suspended from a superstratum. They are especially interested in suspensory behavior and conclude that the orangutan makes the most use of all of its appendages in arboreal situations and, furthermore, shows the greatest variety of suspensory behavior. Here I think we see a speculative solution to the paradox; it is that the very intricacies of his arborealness provided Abang with a number of keys to unlock a strange door.

I conclude from watching Abang that manipulative disability is not a factor that would have prevented apes or Australopithecines from mastering the fundamentals of stone technology.

Mental Capability

Many of my conclusions are woven into the commentary. Here I will discuss one observation for emphasis. Though I have described a case of imitative learning, it would be misleading to think of Abang's behavior solely as imitative learning. To do so would be to miss an adaptive behavioral mechanism that apes possess—random exploratory activity and innovations that can lead to new successes. The motives underlying this behavior are irrelevant to an appreciation of its adaptive value. In other words, I do not have to enter the "insight" controversy. Those behaviors that have adaptive value depend on the experimental environment. In the case of Abang some innovations were valueless in their effect—dipping the flake in water did nothing to improve its effectiveness. Some innovations were neutral replacements; for example, I severed the cord with a cutting action using the flake like a knife. Abang copied this action successfully. Soon, however, he used a chisel action also, pushing the flake against the cord. I could not see any difference in the effectiveness of his innovation over the original imitation.

However, when he detached a tiny chip and put it in his teeth to cut the cord, he demonstrated an innovation of great potential value. It did not actually work as a means of severing the cord, but this was almost certainly because I accidentally discouraged him. He had invented a new technique by himself—the mouth method. Which technique is better? Clearly, if you can open the box with a tiny chip you use less effort and less raw material. Transfer this practice to the wild state and it is not hard to see the potential advantage of Abang's innovation.

Taxonomic Significance

Can the idea of *human* be given scientific legitimacy? Can there be a zoological taxonomic term that is a closer equivalent of the somewhat metaphysical notion *man*? It has at various times been proposed that toolmaking be taken as the rubicon that, when crossed, justifies the application of the term *human* and also whatever official taxonomic classificatory term is devised.

I mentioned above Wells's (1959) classificatory ideas. They led to a supporting note from Robinson (1959), a dissenting one from myself (Wright 1959) and an answer from Robinson (1960). Robinson (1959) proposed retaining the conventional division of the superfamily Hominoidea into two families, Pongidae and Hominidae. The Hominidae he thought could be divided into two subfamilies, the Australopithecinae and Homininae. What is the basis of this last division? Toolmaking is of first importance, a character of the Homininae. As Robinson says, "far greater even than the morphological is the behavioural difference between the two groups. The latter difference involves the matter of culture and represents a major evolutionary step. It cannot be ignored taxonomically" (Robinson 1959:125).

In a rather brashly worded objection to Robinson, I wrote that there were problems for an archaeologist. Java man, for instance, had not been found with tools. More cogently, I expressed some surprise at the legitimacy of including cultural behavior into what I took to be the pure genetic world of zoological taxonomy. Robinson (1960) quite reasonably rebutted some of my points, but one remained unanswered. I did not pursue it because I could not then perceive it clearly.

I think I do now. Robinson maintained (1960:59) that *ethological* (behavioral) *characters* could legitimately be used in taxonomy:

> Behaviour patterns are commonly highly specific and in the final analysis depend on the genetic mechanism of the animal concerned since clearly the behaviour cannot exist apart from the morphology of the creature. An animal's behaviour is an integral part of the organism, as much as is the morphology. [1960:59]

It is this reductionism that evades the question of how learning might spread among such highly intelligent animals as the Hominoidea—a reductionism that implicitly invites us to treat culture *as if* it were genetically carried by early hominids. I think we should be aware, once and for all, that the process of learning among the Hominoidea may have so befogged the paleontological picture that the presence of tools is of no taxonomic value. There was, in other words, far too much nongenetic noise in the system.

CONCLUSION

To prevent misunderstandings I want to stress that (1) my experiment is fundamentally a test of imitative learning and not of insight; (2) I have not sought to prove that australopithecines did in fact make stone tools; and (3) there can be no reconstruction of the precise ethological interaction between australopithecines and more advanced hominids.

What my experiment does do is to make it improbable that australopithecines were prevented from imitating stone flaking by deficiencies in their intelligence or for lack of manipulative skills.

I think that the orangutan's achievements would not have been confidently predicted. Nevertheless, the work does not throw new light on the capabilities of apes, whose exploits under training have for long transcended Abang's simple actions. Rather, the results affect the ranking of stone flaking as an indicator of capability. We can all give patronizing applause to the termiting efforts of Goodall's chimpanzees. We have no difficulty in appreciating the tools they make and what they do with them. Furthermore, no technical language can obscure our mental image of what they are doing.

This is not the case with stone flaking. Very few people have used "a spheroidal hammerstone to direct a blow on the perimeter of a core of siliceous rock to produce a flake by conchoidal fracture." There can now be no confusion; the skill is easy to acquire even at the pongid level.

Acknowledgments

For their assistance with this work I am much indebted to the staff of Bristol Zoo and to Vernon Reynolds, Jonathan Musgrave, and John Crook of the University of Bristol. John Clegg, Les Hiatt, and Peter White of the University of Sydney and Peter Ucko of University College, London, have given me helpful comments on drafts of this article. I thank Derek Freeman of the Australian National University for his encouragement.

References Cited

Broom, R., and J. T. Robinson
    1950 Man contemporaneous with the Swartkrans ape-man. Am. J. Phys. Anthropol. 8:151.

Carpenter, C. R., and N. M. Durham
    1969 A preliminary description of suspensory behavior in nonhuman primates. *In* Recent Advances in Primatology, Vol. 2. H. O. Hofer, ed. Pp. 147–54. Basel, N.Y.: Karger.

Goodall, J. M.
    1965 Chimpanzees of the Gombe Stream Reserve. *In* Primate Behavior; Field Studies of Monkeys and Apes. I. DeVore, ed. Pp. 425–73. New York: Holt, Rinehart and Winston.

Holloway, R. L.
    1968 Cranial capacity and the evolution of the human brain. *In* Culture: Man's Adaptive Dimension. Ashley Montagu, M.F. ed. Pp. 170–96. New York: Oxford University Press.

Inskeep, R. R.
    1969 Some problems relating to the Early Stone Age in South Africa. S. Afr. Archaeol. Bull. 24:174–81.

Leakey, L. S. B. and V. M. Goodall
    1969 Unveiling Man's Origins. London: Methuen.

Leakey, M. D.
    1971 Olduvai Gorge: Vol. 3. Cambridge, Mass.: The University Press.

MacKinnon, J.
    1971 The orang-utan in Sabah today. Oryx 11:141–91.

Napier, J. R.
    1960 Studies of the hands of living primates. Proc. Zool. Soc. London, 134:647–57.

Robinson, J. T.
    1959 Hominid nomenclature. Man 59:125.

    1960 Hominid nomenclature. Man 59:59–60.

Schaller, G. B.
    1965 Behavioral comparisons of the apes. *In* Primate Behavior: Field Studies of Monkeys and Apes. I. DeVore, ed. Pp. 474–81. New York: Holt, Rinehart and Winston.

Tobias, P. V.
    1971 The Brain in Hominid Evolution. New York and London: Columbia University Press.

Wells, L. H.
    1959 'Human', 'hominine', 'hominid'. Man 59:30–1.

Wright, R. V. S.
    1959 Hominid nomenclature. Man 59:217.

Dr. Krogman in the role of Sherlock Holmes.

Wilton Marion Krogman

The Planned Planting of Piltdown: Who? Why?

For over 40 years the image of Piltdown confused our understanding of human evolution. Countless hours were wasted in descriptions and discussion of the "fossil." Krogman summarizes these events from discovery to ultimate solution. He tells us of the individuals he knew personally, of their involvement in the history of this most infamous fabrication, and how the Piltdown forgery influenced the whole interpretation of human evolution. Today, with careful excavation methods, radiometric dating, and the existence of series of fossils, no single specimen could be so influential.

In 1972 Ronald Millar, a professional writer, published a book that he called *The Piltdown Men*.[1] The central character is, of course, Piltdown I, announced triumphantly on December 18, 1912, and devastatingly exposed as a hoax 41 years later.

In his book Millar ranges far afield in discussing and elaborating the climate of human paleontology up to 1912. He goes well beyond 1912 in his references to other human fossils, but such material is almost a non sequitur. The focus, as I see it, is why and how Piltdown I was initially accepted as a legitimate find somewhere along the line of hominid evolution.

239

HISTORICAL BACKGROUND

Victorian-Edwardian times were steeped in the post-Darwinian controversies of creation versus evolution; Cuvier's "catastrophism" was a rationalization of apparent successive geological epochs. In 1891–1892 Dubois uncovered the "ape-man" *Pithecanthropus,* already postulated several decades earlier as a possibility by Haeckel, who called for a pre-Neanderthal *affenmensch.* The confused state of affairs is evidenced by the fact that in 1896 *Pithecanthropus* was labelled anthropoid by five experts, human by seven, and halfway between ape and man by seven. At the turn of the century, Woodward—soon to be deeply involved in the discovery of Piltdown— made some predictions concerning the physical traits of early man, almost certainly influenced by the *Pithecanthropus*-Neanderthal sequential continuum.

There were other strutters on the stage of evolutionary time, some artefactual, others osteological. As early as 1797 Frere reported very early fabricated flints at Hoxne, in England. In 1846 Boucher de Perthes uncovered stone tools associated with the bones of extinct animals at Abbeville, in France. Neanderthal remains were found at Gibraltar in 1848, and at Düsseldorf, Germany, in 1856. In 1856 Lartet of France announced finds of *Dryopithecus* and *Ramapithecus,* placing them in the category now recognized as hominoid. In 1863 a jaw and flint tools were found at Moulin Quignon, in France, in strata said to be as old as those containing fossil bones of extinct animals. Later, it was revealed that the Moulin Quignon site was comparatively modern: the jaw was not old; moreover, the flints had been faked! In 1898 at Galley Hill, in England, skull fragments and a nearly complete skeleton were found, and were stated to represent very early ancestral hominid remains. The find had a curious history, for it was in the private museum of one Robert Elliot, and not until 1908 was it concluded that Galley Hill was similar to the neanderthaloids of Spy, Belgium.[2]

I see no reason to go further with this sort of cataloging. Suffice it to say that by 1912 there was undeniable factual evidence of "fossil man," that is, early and primitively hominid types that to one degree or another attested to a progressive emergence of man from a basic anthropoid ancestry. The real problem, in 1912, was just *how* the bridge was crossed; or to put it in simpler terms, what apelike traits plus manlike traits would combine at a given developmental or evolutionary stage.

There is another state of affairs to be considered in the late 19th and early 20th centuries of England. There was no real science of either archaeology or of human paleontology. These affairs were largely in the hands of amateur antiquarians and natural historians. There were many local societies given over to the unearthing and study of flints and bones, ranging all the way from the *eoliths* of Harrison to the cultural remains of Roman and medieval times. The chairman or president of such a society was often a man of local prominence—lawyer, physician, curate, or squire, who was usually self-taught in the problems of excavation, documentation, and interpretation.

Such a man was Charles Dawson, discoverer of Piltdown. He was a barrister and a leading member of the Sussex Archaeological Society. There is no doubt but that he knew the Sussex area—in which Piltdown was found—quite well, indeed. There is considerable reason to believe that his interests and knowledge did not extend much beyond Sussex, that is, he seems to have been quite provincial in his total outlook.

In addition to Dawson, the principal players in the Piltdown drama were: Arthur Smith Woodward (later Sir Arthur), an ichthyologist concerned principally with fossil fishes; Arthur Keith (later Sir Arthur), anatomist and human paleontologist; Grafton Elliot Smith (later Sir Grafton), neuroanatomist and Egyptologist; Ray Lankester (later Sir Ray), a director of the British Museum; W. P. Pycraft, ornithologist of the British Museum and science writer for *The Illustrated London News;* and Father Teilhard de Chardin, geologist, paleontologist, and philosopher. As excavators, discoverers, and interpreters of the Piltdown remains, these men wrote the script and defined the roles and the scenes.

Let us now turn to the actual findspot at Piltdown. Prior to 1912 a laborer at the site (said to have been gathering gravels with which to repair or construct a stone walkway) reported finding several skull fragments. In 1911 Dawson found a larger fragment of the frontal, presumably related to the earlier finds. In 1912 Woodward joined Dawson, and more fragments were found in a deeper depression in undisturbed gravel; Dawson, alone, found the right half of a mandible, with symphyseal area and condyle broken away; Woodward, working in an area about one yard from the jaw, found a small fragment of an occipital bone. Also found at the site were two small pieces of a molar of an early Pleistocene elephant, a much-rolled cusp of a Mastodon molar, two teeth of a later Pleistocene hippopotamus, and two molars of a Pleistocene beaver. These dental elements encompassed an Early-Late Pleistocene range. On April 13, 1913, Father Teilhard found a canine tooth in an area previously worked over by Dawson and Woodward (who were present when the canine was found). Also found were two articulated nasal bones.[3] Oakley (1973) says that nothing more was ever found at the original site, although the gravels were searched for many years.

In 1915 Dawson sent a postcard to Woodward announcing the finding[4] of parts of a second Piltdown skull. Woodward reported this in 1917, about a year after Dawson's death. Found were skull fragments and a molar tooth. The skull bones, especially a right frontal fragment, were as thick as P I, and the molar was identical with one in the jawbone of P I. At the time, P II seemed to put the clincher on P I. In 1914 a bone implement, 16" long, 4" in diameter, had been found under a hedge at the P I site. It had been fashioned from the femur of a large extinct elephant.

So much for the listing of parts and objects found. The real problem is: what did they represent in terms of time-age, or antiquity, and in terms of evolutionary stage or their exact place in the ape to human progression. One fact is paramount: the way in which skull and jawbone had been broken left an almost incredible variability in reconstructive interpretation. In 1915 Miller wrote as follows: "Deliberate malice could hardly have been more successful than the hazards of deposition in so breaking the fossils as to give free scope to individual judgment in fitting the parts together" (p. 1). One major circumstance emerged, namely, that the skull seemed at one evolutionary stage (human), and the mandible at another (anthropoid).

It is not the place here to go into the controversial details involved in deriving an entire cranial reconstruction from skull fragments with missing contact margins. For example, Keith reconstructed one cranial half and mirror-imaged the other half, assuming symmetry. Smith assailed this, claiming asymmetry, for as a neuroanatomist he followed the dictum of *left cerebral dominance,* with fuller development of the left side of the cranial vault. This sort of reasoning by each man gave considerable difference in the estimation of endocranial volume. The mandible also presented reconstructive and interpretative problems: (1) its symphyseal area was incomplete so that details of the chin development were open to inference; (2) the condyle was missing, so that details of the temporomandibular joint and the hinging of the jaw, with its pattern of chewing, could be only conjectured.

The big question is why men of undoubted ability were able to reconcile skull and jawbone in a single evolutionary form or stage.[5] The answer is found in the 1912 adherence to the ape-man concept, with its implication of a possible differential disparity in evolutionary rate in parts or areas of the skeleton. In substance, it did not really appear anachronistic that craniocerebral evolution proceeded at a pace faster than, for example, dentomandibular evolution.

Indeed, as late as 1925 Hooton wrote a paper on the asymmetric character of human evolution. On a 1–6 ordination he assessed a number of nonmetric craniofacial and dental traits, from ultraanthropoid to ultrahuman. He rated the Piltdown skull 4–5 in the skull (inferior human/typically human); and 3–4 in palate (subhuman, supraanthropoid/inferior human); and 2 in mandible and teeth (typically anthropoid).[6]

It seems to me that this type of thinking regarding the uneven or asymmetric pace of evolution tended to quell doubt and inhibit challenge as to the human/anthropoid traits in P I. "This is the way evolution proceeds: at different rates within the craniofacial complex" appeared to be the ruling doctrine. And if this doctrine were adhered to by men of authority and competence, who is to gainsay?

Time and technique catch up with one another. By 1950 two factors were operative: (1) new and more precise methods of dating fossil material; (2) a generation of men not only trained in newer methods, but also not afraid of challenging authority.

Criticism began to focus upon three main areas: (1) the antiquity or age of Piltdown; (2) the color or staining of bones and flints; and (3) the details of bone and tooth morphology, especially the latter.

In 1943 a Dr. Lewis, a researcher of the Imperial Chemical Industry, gave a lecture on the relationship between subsoil and health. In it he referred to the work of Owen Rees (c. 1850) on the fluorine (Fl) content of bones. In 1873 Carnot followed up Rees's work and gave the proportion of fluorapatite in bones as follows: modern = 0.058 percent; Pleistocene = 0.33 percent; Tertiary = 0.62 percent; Mesozoic = 0.91 percent; Paleozoic = 0.99 percent.

In 1949 Oakley pointed out that later Pleistocene fossil bones had less Fl than early Pleistocene. He gave a range of 0.1–0.5 percent for late Pleistocene, 1.9–3.1 percent for early. The Piltdown I skull, jawbone, and canine had very little Fl ($<$ 0.1–0.4 percent). In 1955 Weiner gave P I skull = 0.1 percent, mandible = 0.3 percent; in P II skull = 0.1 percent, molar = 0.01 percent. Weiner also said that the N (nitrogen content) test was confirmatory. Finally, in 1973 Oakley gave Carbon 14 ages in P I as follows: skull fragments = 1330 A.D. (620 +/− 100 BP Gr N − 2203); jawbone 1450 A.D. (500 +/− 100 BP Gr N − 2204).

The verdict is clear: P I (and P II) are modern as far as antiquity is concerned. "None of the [Piltdown] fragments can be considered fossil or in any way part of man's ancestry" (Oakley 1973:6).

The gravel at Piltdown and the bones and teeth of P I and P II, as well as the flint tools, had a reddish brown color, presumably due to oxidation. At first the color of the bones was explained by saying that Dawson had dipped them in dichromate of potash "to harden them." It has become obvious that bones, teeth, and flints had been intentionally stained to match the color of the gravels in which they were apparently deposited: the skull fragments were dipped in potassium dichromate; the jawbone color was due in part to Fe staining plus large traces of chromium; the canine had a dark coating attributed to iron oxide, which later proved to be Vandyke brown paint.

The skull fragments had a high content of gypsum ($CaSO_4 . 2H_2O$) that replaced much of the apatite ($Ca_5 (PO_4)_3 OH$). Oakley (1973) states, based on experimentation, that bones soaked in iron sulphate will have much of their apatite converted into gypsum. Up to 2 percent of chromium was found in the stains of the fossil teeth, on one animal bone, and on one flint, in spite of the fact that the Piltdown gravels contained no comparable amount of chromium. "It seems as though the "hoaxer" used chromium compounds to assist the oxidation of the iron salts with which he was staining specimens to match the reddish-brown gravel of Piltdown" (Oakley 1973:4).[7]

It may be apropos at this point to discuss the bone implement, as well as other fossil animal bones. The bone implement was pointed at one end and rounded at the other by a series of cleanly cut facets. Careful study demonstrated that the facets were *not* cut by a flint tool, but were the result of the action of a steel knife on fossilized bone.

The fossil material was not of Piltdown provenience. Millar (1972) says that an elephant molar *(E. planifrons)* came from Tunisia, and that a hippo tooth came from Malta or Sicily. Weiner specified that the *E. planifrons* molar was traced to Ichkeul, in Tunisia near Bizerte. Of 18 fossil specimens, says Weiner, 10 were fake, eight probably fake.

Piltdown II was very clumsily contrived, apart from the fact that its find spot was never determined. The right frontal fragment belonged to P I; the frontal and occipital fragments were also artificially stained, and the two pieces did not even belong to the same skull. The molar was obviously a leftover of the P I mandible; the fragments were associated with a Red Crag rhino tooth, an utterly impossible association. As confirmation of P I, the P II material was an utter flop!

The canine tooth found in 1913 by Father Teilhard was smaller than that of an ape, but larger than that of a man. It was, in truth, just about the size that Woodward, a decade or so earlier, had predicted. It was greatly worn, deep into the pulp cavity. A short while after it was found, it was X-rayed by a dentist named Lyne. He said at once that it was an immature tooth—perhaps newly erupted—for there was no apical closure. Its amount of wear was completely at odds with its evident immaturity. Of the mandibular teeth as a whole Keith said only that the roots were human in type.

The right jawbone, carefully broken in the midline and in the condylar area, artfully removes details of chin and temporomandibular structuring; one cannot, from the jawbone, determine whether the bite was up-and-down or rotatory or grinding. The chewing surfaces of the teeth are worn flat, as one might expect in the tough, gritty diet of an early or primitive hominid. The flatness was achieved by artificial abrasion, that is, by grinding. This is especially true for the P I canine and the P I, II molars.

The jawbone has now been identified as that of a young female orangutan. I studied the jawbone at the Royal College of Surgeons, London, in 1931, and concluded it was that of a chimpanzee. I made this decision because the wear suggested that rounded or bunodont cusps had been worn down. I did not suspect orangutan, for the crenulated molar cusp pattern of the occlusal surface of the orangutan molar was not in evidence; and no wonder, for all such crenulations had been carefully removed by abrasion to give a false human wear-pattern.

And so we come to the end of the reviewing of a very carefully and artfully conceived and achieved hoax. As I review the unfolding of the details of the fakery I feel that the perpetrator made only one major mistake, that is, the creation of a second Piltdown. It was done originally, perhaps, to shore up whatever doubts P I may have raised. To have *two* sets of skeletal remains of an early Englishman was to say, in effect, lightning *did* strike twice! The involvement of P II with bones and teeth of P I, and with bone fragments of two different skulls, did not gild the lily—it took the bloom off entirely. In essence, what at first seemed a good thing was pushed too far, pushed over the brink of credulity.

With all the evidence of mischievous and intentional chicanery before us, it is logical that we turn to our first question: *Who?* We have noted the principals in this affair, those who were, in one way or another, directly involved. There is one obvious man to point to first, Charles Dawson. The whole thing was on his own home grounds, so to speak. If opportunity be a prime factor, then it is Dawson to whom we must look. He was a known and established antiquarian who collected flints and other early artefacts, as well as historic material (Romano-British, early Norman times). Also, he had a known knowledge of the use of various chemicals to harden flints or other more friable artefacts. Obviousness, however, can be a slender reed of support, and so, for the moment, let us pass on to the others.

Sir Arthur Smith Woodward was the man to whom Dawson most often turned. It is certain that he spent short periods of time at Piltdown, digging and sorting through the gravel deposits from whence the bones and teeth assertedly came. But it seems that he came down to Sussex only upon call by Dawson. None of the contemporary reports or articles on Piltdown suggests his presence on the scene on an a priori basis, that is, for any protracted period prior to a dig or before a new find was announced. In other words, opportunity for Woodward to salt the area was negligible. Apart from this sort of mechanics of cheating, it must be recorded that Woodward's record of scientific probity and professional ability was of the highest order.

Father Teilhard de Chardin found the canine tooth on August 30, 1913. Because of his status as a paleontologist-geologist, he was, we are told, invited by Dawson to participate in looking through the Piltdown gravels. On that day Dawson and Woodward were also present, digging and sifting. Suddenly, in an area of gravel known to have been recently gone over, the canine was found. It is recorded that the two men (Dawson and Woodward) were well-nigh incredulous: the tooth seemingly had come from nowhere. Father Teilhard was also apparently involved in the matter of the tooth of *E. planifrons,* traced to Ichkeul in Tunisia. It became known that he had been to Ichkeul shortly before coming to the Piltdown site. Because of this combination of circumstances, there were those who pointed the finger of suspicion at Father Teilhard. This was the first find of *E. planifrons* ever made in England; not until 1920 was another such find made.

In later years de Chardin did not refer to Piltdown in his discussion of fossil man. In a 1952 article in a publication of the N.Y. Academy of Sciences, he shows *Eo.* as middle Pleistocene in an evolutionary diagram.[8] He goes no further than this notation. One may conjecture as to the reason for this failure to enlarge upon Piltdown and his role in the discovery. Perhaps residual doubts as to authenticity had crystallized into a feeling that it was better to let sleeping dogs lie than to come forth with a denunciation that might weaken the scientific validity of Darwinism and the emergence of man from an ancestral form. I am inclined to this interpretation, for I later met Father Teilhard, although I did not get to know him well. As I heard him speak and as I read his later books, I gained an impression of a spiritual devotion to evolutionary concepts and a literally universal factor (design) in the unfolding of the entire cosmos. I simply cannot see this man as the perpetrator of a fraud.

It was my privilege to study with Sir Arthur Keith as a National Research Council Fellow, 1930–1931, at the Royal College of Surgeons, London. We discussed problems involved in the reconstruction of the Piltdown skull and my conclusion that the mandible did not belong to the skull and that it was chimpanzoid. As far as I know, Sir Arthur did not visit the Piltdown findspot. He never mentioned doing so to me and I know of no reference by either Dawson or Woodward that he had done so. Sir Arthur was a comparative morphologist (infrahuman primates and man) to whom finds were referred for an opinion. He was in principle and in fact a stay-at-home museum man rather than a field man. If in the early days he was taken in by the Piltdown bones and teeth, it was in part due to the then-believed acceptance of disparate rates of evolution, plus confidence in, and loyalty to, his colleagues, notably Woodward.

While I was studying with Sir Arthur I had the opportunity to meet Sir Grafton Elliot Smith, in 1931. He had just returned from Peking (Choukoutien) where *Sinanthropus* crania were being unearthed. He had made some endocranial casts of the *Sinanthropus* brain, a subject in which I was interested. As a new young Ph.D. in physical anthropology, I wanted to learn all I could about endocasting and the light it could throw upon cerebral evolution and development. I had studied neuroanatomy at Chicago with C. Judson Herrick and had read many of Sir Grafton's publications in this field. I did not talk with him about Piltdown; frankly, I shied away from possible involvement in the Keith-Smith controversy over the reconstruction of the Piltdown skull, for I had heard that Smith could be rough-tongued.

In my study with Sir Arthur I was more interested in modern racial craniology than I was in human paleontology. Yet since Keith's books on the subject had figured prominently in my graduate studies, I often discussed this or that fossil skull or other bone with him. As we discussed Piltdown he always referred to differences of opinion between him and Sir Grafton quite mildly and always recognized one's right to disagree. I formed the very definite opinion that Sir Grafton, like Keith himself, had a more academic professional stake in Piltdown rather than a more personal interest with reference to the finding itself and the circumstances thereof. At no time did Sir Arthur mention that Sir Grafton had visited the findspot, much less search the gravels as Dawson, Woodward, and de Chardin had done.

W. J. Pycraft was Curator of Anthropology under Sir Arthur Smith Woodward. I met him in 1931 at the Kensington Branch of the British Museum, when he made it possible for me to see the skull of Rhodesian Man.[9] Pycraft was an ornithologist, but at Woodward's request he studied the Piltdown jaw and teeth. He wrote to Gerrit Miller that they were human, seen grossly and radiographically, and were not anthropoid. There is no record that I know of that suggests active field (findspot) participation. Apparently he entered the Piltdown picture only upon request of his chief, Woodward.

Sir Ray Lankester, though he early doubted that the jawbone belonged with the skull, did have an interest in the Piltdown find. He was an ardent champion of *eoliths* or *dawn-stones,* claimed to be the very earliest intentionally but crudely fashioned flint tools. He saw in Piltdown the maker of rostrocarinate flints (a type of eolith). Moreover, he scouted the possibility that Piltdown might authenticate Ipswich man, recently found by J. Reid Moir, a foremost protagonist of the eolith theory. Whether Sir Ray had any real contact with the Piltdown site I cannot ascertain.

As the previous few pages are scanned in reference to the question *Who?* I suggest a box score as follows:

No	*Possibly*
Woodward	Dawson
de Chardin	Smith
Keith	
Pycraft	
Lankester	

I venture this dichotomization for two reasons: (1) a rather generally accepted view that Dawson is a candidate because of his on-the-grounds relationship to all circumstances of the find; and (2) the fact that Millar specifically regards Smith as having the necessary anatomical knowledge and, perhaps, the selfish motivation to use Piltdown for personal advancement.

What I propose to do here is to give the pros and cons for each man and then to weigh them as objectively as I can. I shall start with Dawson.

Evidence for Dawson

1. The entire hoax was very complicated, for it seems to have demanded considerable sophistication in skeletal morphology, odontology, paleontology, and chemistry. The entire craniodental ensemble had to be rendered a reasonably acceptable whole, not only within itself but also with reference to a type-stage of evolution. Did Dawson possess the knowledge and the insight to plan all this?

2. The very obviousness of Dawson as a perpetrator tends to mitigate against his being so. Were he shrewd enough to plan the whole thing, he must have known he would be a prime suspect.

3. Hence, on both technical and logical grounds, a defense of Dawson may be raised. In Millar's words he simply "does not fill the bill."

Evidence against Dawson

1. Dawson had an earlier record of deliberate deceit. His fellow members of the Sussex Archaeological Society had accused him of faking flint implement-types. A prickspur he claimed to be Norman was not so recognized by the British Museum. A Roman cast-iron statuette found in a slag heap at a Roman ironworking site at Beaufort, near Hastings, was also suspect. On this point Downes of the University of Birmingham wrote: "We must wait for a complete review of the evidence of Mr. Dawson's activities, ranging from his honest and talented work to his undoubted deceptions, before passing any judgment on him or on such debatable specimens."

2. Sussex was Dawson's own provenience. Of all persons involved, he had the best knowledge and the least restricted opportunity to plant or to salt the Piltdown gravels with faked bones, teeth, and flints.

3. He was known to use chemicals (ferric and chromate) ostensibly to harden or toughen various specimens of bone and flint. By observation alone, various degrees of staining could be noted, learned, and repeated.

4. Dawson's *History of Hastings Castle* was denounced as a plagiarism of an earlier manuscript by William Herbert. It is true that Dawson acknowledged his indebtedness to Herbert's manuscript in a foreword, but it was still felt that this was not enough to excuse very extensive verbatim copying. A slight restatement here and amplification there does not constitute claim to original work.

5. He made absolutely no data available on the circumstance or site of the finding of P II. All that is known is that it was from a field about two miles from the site of P I.

6. In 1917, after Dawson's death, the British Museum received from Mrs. Dawson fragments from what was labelled the Barcombe Mills skull. The assumption is that Dawson had found them, though he did not so report. The bones were stained the same color as were those of P I and P II. They were subsequently revealed as belonging to three modern individuals.

Evidence for Smith

1. The almost unbelievable human bone and tooth assemblage of P II and the presence of an impossibly associated rhino tooth are not the sort of random blunders that a trained anatomist would perpetrate. In his 1904 report on Nubian skeletal material, Smith demonstrated a superb knowledge of osteology. It is neither reasonable nor logical to tag him as a hoaxer.

2. Smith's neurological studies, which were not only comparative but phylogenetic as well (via endocasts of fossil vertebrate crania), would not have resulted in a melange of unrelated human cranial fragments or out-of-place (timewise) fossil remains, including teeth.

3. Dawson was a local barrister, Smith already an admitted authority in the total field of neuroatomy. He had far more to lose than did Dawson were he to be revealed as scientifically unethical and dishonest. Of the two, Smith would perceive the more readily that the P I and especially the P II assemblages were so clumsily contrived as to literally demand disclosure. As fate would have it, the whole hoax was perpetrated in the lifetimes of both Dawson and Smith. The thought occurs to me that one of the two lived in deadly fear for some years.

Evidence against Smith

1. Smith was an Australian. At the time of the Piltdown discovery he had a post in Cairo that was not as prestigious as one in England. It has been suggested that a fossil find such as Piltdown might enhance his status and so lead to advancement to a post in England.

2. Smith was an ardent advocate of cultural diffusion. Indeed, he was a participant in the idea of a *heliolithic wave* of civilization from Egypt; wherever sun worship was found, wherever pyramids or certain other megalithic monuments were found, there was evidence that Egypt was a focal diffusion point. Might not the discovery of early man in England—after many other such discoveries on the Continent or even in Indonesia—be confirmatory evidence of an early, basic hominid centrifugal migration?

3. Did Smith try to put one over when he stated that a human skull and a simian jaw were not incongruous if found in the same individual? Here I shall take up the cudgel for Smith: (1) he but affirmed the current belief (1912) in the possibility of disparate rates of evolution; (2) in his earlier papers on evolution he had—and this is to be expected of a neuroanatomist—held that cerebral evolution

took precedence over even the facial skeleton, not to mention the infracranial skeleton. He reasoned that cerebral development came before the erect posture and the freeing of forelimbs to become toolmakers.[10]

4. Smith was said to be arrogant (Millar), and it might be that he wanted, via hoaxing, to discredit the British Museum pundits. I found Smith forthright and outspoken and have known other scientists who were the same. This, however, is no qualification for dishonorable behavior.

5. Millar cites Smith's 1937 obituary as saying he had a "childlike simplicity of approach to scientific trust." I cannot reconcile arrogance and childlike simplicity!

6. It is felt by Millar that Smith's "complete failure to assist Woodward (in the reconstruction of the Piltdown skull) is in my opinion highly incriminatory." Maybe he wasn't asked to.

7. In 1915 the Talgai skull was found in Australia. It is alleged that Smith told Dawson that it was found at Pilton (which Millar says does not exist). Hence, Talgai should be identified by a findspot of Darlington Downs, to avoid confusion between Pilton and Piltdown. I mentioned this to Raymond Dart recently, and he doubts the whole thing.

8. The C^{14} age of P I dates to medieval times. Millar feels that Smith, rather than Dawson, would have readier access to skulls of some 500 years ago.

Here I end *Who?* and turn to *Why?*

Obviously the vagaries of human motivation, human drive, human ambitions, and human sense of right and wrong are too diverse, too complexly interrelated, to say of a given man, "he's honest," or of another, "he's dishonest."

One cannot particularize; one can only generalize.

1. A chauvinistic national pride may have been a major factor: for England and St. George. Perhaps the hoaxer thought that "there'll always be an England" was not enough, and added, "there always *was* an Englishman." Also, up to 1912 England had no entry in the Early Man Sweepstakes, so, up Piltdown!

2. Personal recognition and fame would certainly come to the discoverer of Piltdown as he vaulted into public and scientific prominence; in those days it might bring him an F.R.S., a knighthood with a royal audience, and a niche in English history.

3. Or perhaps mere professional advancement rather than glory could have been an impelling force, a risky rung on the ladder of achievement.

4. Was revenge a motive, to pay back a real or fancied slight? Was the planning and planting of Piltdown (both I and II) conceived in a cunning, paranoiac mind in order to vaunt self over associates and colleagues?

5. Then, of course, there is always a practical joker in the crowd. Perhaps the hoaxer at first figured to be uncovered, then to excuse himself with an all-in-fun bravado. But it was not uncovered and it was soon out of hand. A bolstering touch here, another there, and then the hoaxer piled the Pelion of P II on the Ossa of P I. Deception fed, cannibalistically, upon itself.

If I now evaluate for and against arguments, I must conclude that the evidence against Dawson is stronger than that against Smith. This view is strengthened as each man is viewed as a person and as a professional, lawyer and trained scientist, respectively. Dawson, the provincial, was relatively supreme on the Piltdown homeground. Smith was the more cosmopolitan outsider who functioned mainly as interpreter of the find. In the early part of this century Dawson lived in a tight little small-town world—a world in which personal approval or public rejection was of paramount importance. Dawson had had his share of criticism and doubt. It could well be that he decided to call the tune to which his fellow Sussex archaeologists must dance, and for 40 years the world of human paleontology danced with them!

Notes

1. St. Martin's Press, New York, 264 pp., 15 plates, six diagrams, two tables. I previously reviewed the book in the *Bulletin of the N.Y. Academy of Medicine,* 19(11): 1011–1016, Nov. 1973. The present article is an amplification of that review, with the kind permission of Saul Jarcho, M.D., Editor-in-Chief of the *Bulletin.*

2. In 1949 Oakley found Galley Hill to be recent and "modern," based on fluorine dating; skull = 0.3%, mandible = 0.4%, right tibia = 0.4%, a limb bone fragment = 0.4%, left femur = 0.2%.

3. Millar (p. 133) writes thusly: "Two nasal (turbinal) bones were found by Dawson. . . ." They were *not* turbinal bones, which are much too fragile to have survived a water-borne deposit. They were the right and left nasal bones of the bony nasal skeleton. Keith thought these nasal bones were australoid-tasmanoid, i.e., short and broad.

4. Oakley (1973) said that later it proved impossible to identify the field for further research.

5. Even though Gerrit Miller (1915) ascribed the jaw to a new species of chimpanzee *(Pan vetus).*

6. More specifically, Hooton gave brain-case 4.31, face 2.30, (total 3.63, S.D. 1.17).

7. Weiner (1955) suggests also that ferric ammonium sulphate (iron alum) was used as a staining agent.

8. The *Eo.* is for *Eoanthropus dawsoni,* the scientific term given to Piltdown. I am indebted to Gabriel Lasker of Wayne State University (Detroit, Mich.) for calling this reference to my attention.

9. The pelvic bone of Rhodesian man was a sore subject with Pycraft. In a completely incorrect manner he reoriented the acetabulum (hip socket) in order to produce flexion at the hip. He dubbed the result *Homo cyphanthropus,* or stooping man. Needless to say, his reconstruction was vigorously criticized.

10. We now know that this is not so. The australopithecines achieved erect posture and freed the forelimbs before the brain developed its full human specialization.

References Cited

Hooton, E. A.
    1925   The asymmetrical character of human evolution. Am. J. Phys. Anthropol. 8(2):125–141.

Keith, Sir Arthur
    1929   The Antiquity of Man. Vol. II, pp. 486 ff. London: Williams and Norgate.

Millar, Ronald
    1972   The Piltdown Man. New York: St. Martin's Press.

Miller, G. S., Jr.
    1915   The jaw of the Piltdown man. Smithson. Misc. Coll. 65(12):1–31.

    1918   The Piltdown jaw. Am. J. Phys. Anthropol. 1(1):25–52.

Montagu, M. F. A.
    1951   The Barcombe Mills cranial remains. Am. J. Phys. Anthropol. N.S. 9(4):417–426.

Oakley, K. P.
    1950   The Fluorine-Dating Method. Yearbook Phys. Anthropol., 1949, pp. 44–52.

    1973   The Piltdown man hoax. Paleontology Leaflet #2. London: British Museum (Natural History).

Oakley, K. P. and C. P. Groves
    1970   Piltdown man: the realization of fraudulence. Science, Aug. 21, p. 789.

Smith, Sir Grafton
    1907   Asymmetry of brain and skull. J. Anat. Physiol. 41:236–247.

    1927   Essays on the Evolution of Man. Second ed. London: Oxford University Press.

    1929   Migrations of Early Culture (Second ed.). *U. Manchester Ethnol. Series* #1. Manchester: University Press.

Smith, Sir Grafton and F. Wood Jones
    1910   The Archaeological Survey of Nubia 1907/08. Vol. II, Report on the Human Remains. Cairo, Egypt.

Weiner, T. S.
    1955   The Piltdown Forgery. London: Oxford University Press.

Dr. Schultz being groomed by Minnie Macaque.

Adolph H. Schultz

Illustrations of the Relation between Primate Ontogeny and Phylogeny

For many years Adolph Schultz provided data and insights on the primates in numerous major papers and illustrated his work with his own extraordinary drawings. The following pages include many of these pictures. We hope that this "Schultz Atlas" will direct attention to the excellence of his pictorial portrayal of the primates.

INTRODUCTION

The following "Schultz Atlas" consists of old and new drawings the author selected to illustrate some of the ontogenetic changes and corresponding phylogenetic specializations in primates. These two broad fields of interest are inseparable because the widely differing degrees of morphological alterations during individual development are initial causes not only for evolutionary innovations, but also for intraspecific variability, including secondary sex differences. With the addition of a few examples of abnormal conditions, it has here been merely indicated that certain congenital disturbances in ontogenetic processes also may lead to the appearance of taxonomically new characters. Because the combination of figures, chosen from the large accumulation of the author's drawings, seems to be very heterogeneous, all these drawings had to be accompanied by brief explanatory notes in regard to their bearing on the above-mentioned major interests. Full discussions of the various studies for which these drawings had originally been intended can be found in the writer's publications, referred to in parentheses. Some of these illustrations have already become well known after repeated use in various recent books.

Modern anthropology has resulted from man's unique interest in his own nature and has gained most through systematic comparisons, especially between man himself and his contemporary simian relations, between man's present condition and that of his antecedents, and between different stages of human growth. Long ago the author had tried to emphasize these facts in an ex libris (see Figure 1).

ONTOGENY

The venerable recapitulation theory, according to which ontogeny supposedly recapitulates phylogeny in abbreviated form, has gradually been replaced by the realization that the same original processes of individual

FIGURE 1
Allegory of man's study of his own nature by means of comparisons, showing age changes in body proportions, evolutionary changes in the head, and differences between the formation and posture of ape and man.

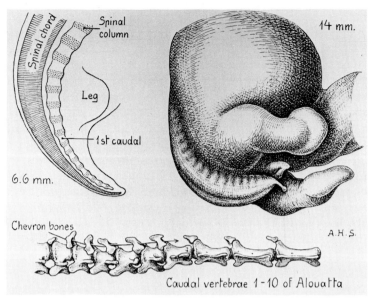

FIGURE 2
Diagrammatic sagittal section through caudal region of a human embryo of 6.6
mm. crown-rump length and caudal part of an older human embryo of 14 mm.
crown-rump length, showing the external tail with caudal filament and the
inverted broad feet with as yet no digital separation. *Below:* proximal tail vertebrae
with their chevron bones of an adult howler monkey. (after Schultz 1969a).

development are tenaciously retained during evolution except for any
changes that have been connected with the appearance of new taxonomic
characters. Among others, De Beer (1930) had demonstrated this with a
profusion of supporting evidence and concluded that phylogeny is due to
modified ontogeny. The same general idea underlies Bolk's (1926) so-called
fetalization theory, which claims that human evolution resulted from a
retention of fetal conditions through ontogenetic retardation. This, however, can be maintained in regard to only a very limited number of characters, since in many other respects developmental specializations of man are
clearly accelerated in contrast to corresponding conditions in nonhuman
primates (Schultz 1926, 1956a, Starck 1962). Some of these facts are
shown by the following examples.

The repetition of ancestral ontogenetic processes during evolution is
clearly illustrated by the development of an outer tail in embryos of man and
of all other *tailless* primates. As shown by Figure 2, early in prenatal life the
human tail reaches far below the initial leg-buds and contains more than
twice as many segments as persist later on when this outer tail soon becomes
overgrown by the surrounding structures. Furthermore, even in man so-
called chevron bones still develop regularly on the ventral side of the first

few tail vertebrae, where they straddle the main caudal artery throughout life in longtailed monkeys, while in man they are resorbed, except in rare cases of persistence to adulthood (Schultz 1941, 1961, 1969a). At a slightly later stage of development the outer tail has already become proportionately much smaller in man, whereas considerably longer in a rhesus monkey, as shown in Figure 3. In most other respects, however, these two primate fetuses are still strikingly alike due to similar rates of early growth of the head, trunk and limbs. This is clearly evident, e.g., in the feet, which become so widely different during the subsequent development of monkey and of man. To demonstrate the profound ontogenetic changes in the human hand and foot the author had modelled and enlarged these structures of a human fetus of seven weeks to bring the early foot to the same length as his own adult foot of which a cast is shown alongside in Figure 4. It is readily seen that hand and foot not only increase more in length than in breadth, but also change markedly in the proportionate size of their parts. The first toe of the human fetus is still relatively short and abduced, while

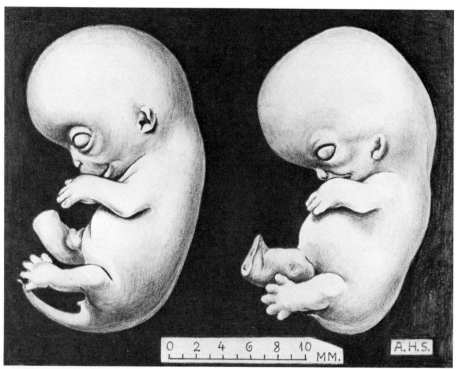

FIGURE 3
Fetus of a rhesus macaque monkey of the known age of 44 days and a human fetus of ca. 49 days (after Schultz 1969a).

258

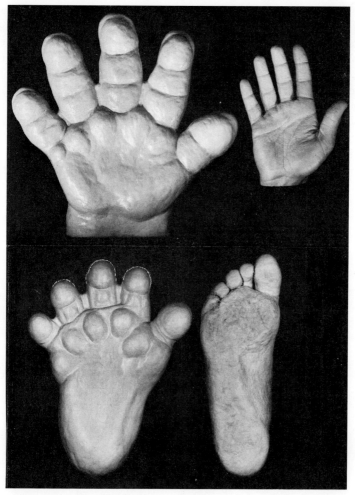

FIGURE 4
Hand and foot of a human fetus of seven weeks (modelled 100 x enlarged) compared with casts of hand and foot of adult man.

the middle toe projects beyond the others and the thumb branches from the base of the second finger without as yet even a trace of its later rotation. The touch pads on the palm and sole are well developed, and so is the early interdigital webbing that still remains from the preceding lack of an outer digital separation. All these conditions are very similar in early fetuses of monkeys and apes (Keibel 1906, Schultz 1956a), but subsequently they undergo very different ontogenetic changes. For instance, with advancing age the touch pads disappear more completely in man and manlike apes than in the lower primates and the thumb does not shift nearly as far proximally, nor does it become as much rotated in platyrrhines as in catarrhines (Figure 5). The phalangeal parts of the toes II to V increase enormously in length in gibbons and orang-utans, whereas those of man become relatively shorter during postembryonic growth (Figure 6) and the webbing between the toes II and III normally persists in siamangs (Figure 7 C) and rarely also in chimpanzees and man. Further examples of phylogenetic innovations in the hands and feet of primates, due to ontogenetic modifications, are shown in Figure 7. The thumbs of spider monkeys and of guerezas have become reduced to mere vestiges, usually embedded in the palm, but occasionally they still sprout a useless outer digit as in figure 7 A. In small fetuses of *Colobus* the author (1924) found such rudimentary thumbs to be less reduced than in adults. In pottos (Figure 7 B) and even more so in *Arctocebus* it is the second finger that has become stunted through localized

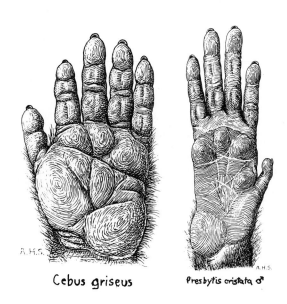

Cebus griseus   Presbytis cristata ♂

FIGURE 5
Hand of an adult capuchin monkey and of an adult langur.

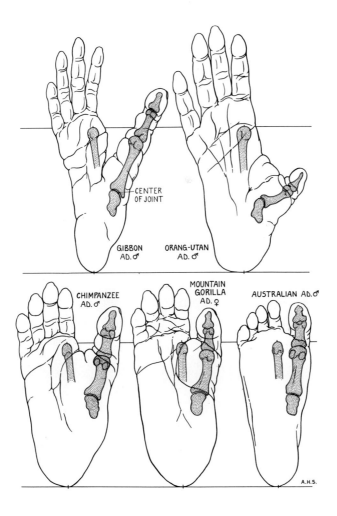

FIGURE 6
Feet of adult hominoids, showing the different relative lengths of the toes (after Schultz 1950a).

261

inhibition of its growth and in over half of all orang-utans the terminal phalanx of the first toe has failed to develop (Schultz 1968) (Figure 7 *D*). In surprisingly many human beings the middle phalanx of the fifth toe is congenitally fused with the distal phalanx or totally lacking (Figure 7 *E*). Extreme changes in digital growth have also evolved in the aye-aye, whose third and fourth fingers reach an enormous length, surpassing even the length of the forearm and that of the upper arm, but affecting the various segments and their thickness in widely differing degrees (Schultz 1954), as shown by Figure 8.

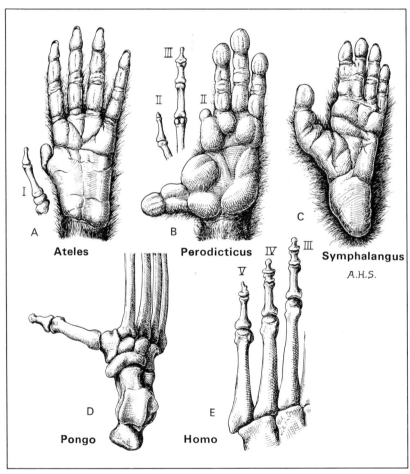

FIGURE 7
Some digital conditions of adult primates, abnormal in most species, but common in the examples shown (after Schultz 1956b).

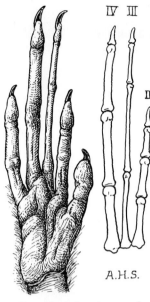

IV III

II

A.H.S.

Daubentonia ad.

FIGURE 8
Hand of an adult aye-aye and the skeleton
of 3 digits of the opposite hand, showing
the extreme elongation of the 3rd and 4th
fingers.

BODY PROPORTIONS

The gradual intrauterine changes in the body proportions of a macaque are readily recognized by a comparison between Figures 3 and 9. In the developing skeleton in the latter figure, typical of catarrhine monkeys, the relatively large neurocranium, the also very large orbit and the central position of the craniovertebral joint are here especially noteworthy as features which persist well-marked to birth (Figure 10). The fateful termination of prenatal life must necessarily occur before the fetus has become too large to pass the maternal birth canal. This depends among most primates chiefly upon the size of the head, which is comparatively huge on account of the precocious growth of the brain. That usually there is little room to spare between the neonatal head and the pelvic passage is very evident from the example in Figure 10 and has been determined in detail for many primates (Schultz 1949, 1961, Leutenegger 1973). The great apes form an exception to this rule by having amply wide pelves and remarkably small relative birth weights, as well as being born in a less mature stage of development than monkeys, wherein they equal man (Schultz 1956a).

In regard to their main body proportions, the simian primates have acquired many clear differences at birth (Figure 11), but in general these are

263

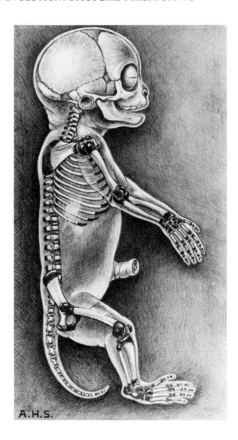

FIGURE 9
Cleared fetus of a rhesus macaque monkey from the middle of prenatal life (after Schultz 1969a).

not yet nearly as pronounced as they become at the completion of postnatal growth (Schultz 1933, 1956a, 1973b). It is of particular interest that the great length of the arms of gibbons and orang-utans is already very evident in early infants in contrast to the human specialization of an extreme leg length that develops ontogenetically late. The postnatal age changes in body proportions are on an average even somewhat greater in man than in the other hominoids and certainly show no unusually close retention of fetal conditions. Especially the relations in length between the front- and hindlimbs to each other as well as to the trunk, change after birth far less in the apes than in man (Schultz 1973b). The skeleton of an adult chimpanzee in Figure 12 reveals many other hominoid characters that have appeared gradually during ontogeny, such as the wide thorax on which the scapulae have shifted far dorsally and the more or less marked promontory at the lumbosacral border. The pongid specializations of an extremely short lumbar region and uniquely long hipbones are clearly evident from a compari-

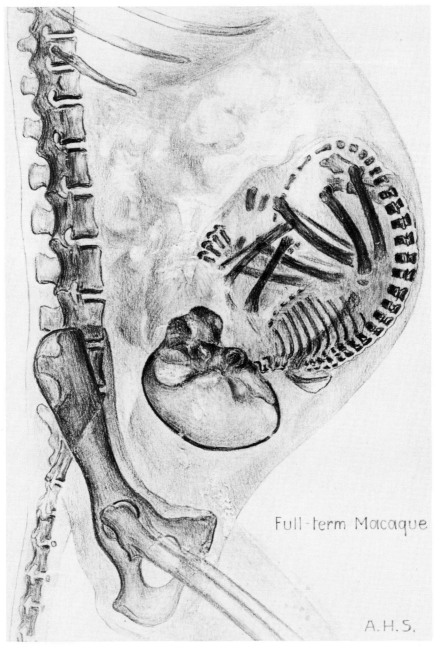

FIGURE 10
X-ray photograph of a rhesus macaque monkey and the pelvic region of its mother shortly before parturition (after Schultz 1969).

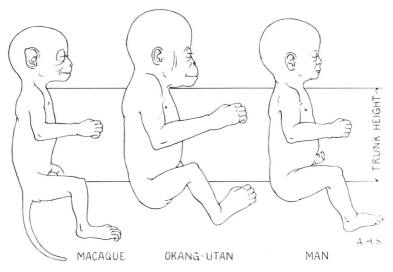

FIGURE 11
Exact side views of some newborns, reduced to same trunk length, showing their different body proportions (after Schultz 1926).

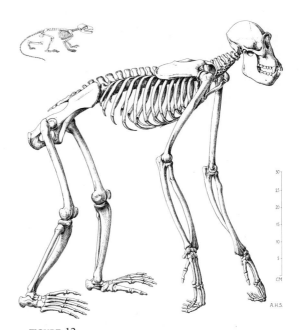

FIGURE 12
Skeleton of an adult male chimpanzee and for comparison of an adult tree shrew, reduced to same scale (after Schultz 1969b).

266

son with the primitive conditions of the adult tree shrew in the same figure or with the two prosimians in Figure 13. The latter possess the relatively shortest and longest hind limbs among primates, the tarsier surpassing even man in this respect. The extreme shortness of the hindlimbs of the gorilla-sized *Megaladapis* becomes particularly impressive by the comparison with the femur of a full-grown *Propithecus* in Figure 14, but the femur of the former had to become many times thicker to carry such a huge body.

It is not only in the main proportions of the trunk and limbs that different rates of growth, together with their duration, produce widely diverging specializations among adult primates and even in the closely related hominoids, but they are also responsible for the gradual appearance of innumerable other distinguishing features, such as the relative position of the head and shoulders and the formation of the neck. The sketches in Figure 15 are intended to show this with one example. In the typical quadrupedal posture of an adult gorilla the bulky trunk is slung between the long arms and short legs while the large face hangs far below the high

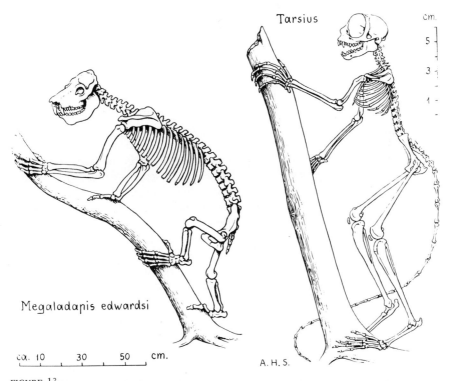

FIGURE 13
The relatively shortest and longest hind limbs among primates in an adult *Megaladapis* and an adult *Tarsius* (after Schultz 1969a).

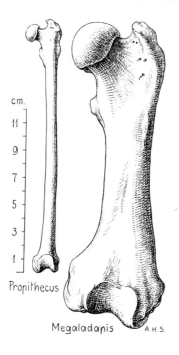

cm.
11
9
7
5
3
1

Propithecus

Megaladapis   A.H.S.

FIGURE 14
Femur of an adult *Propithecus* and an adult
Megaladapis, reduced to same scale (after Schultz
1969a).

shoulders. A similar pose has become very awkward for an adult bipedal human, whose slender neck has to be sharply dorsoflexed to raise the head for horizontal vision and who has to get on his knees to keep the trunk level. A bipedal, fully erect posture, however, is quite natural for even adult orang-utans when the arms are required for carrying, as shown by Figure 16. The lack of a real outer neck in adult gorillas and orang-utans is due to the high position of the shoulders and the low one of the large face, while dorsally it is hidden by the powerful nuchal musculature, needed to balance the head on its joint which ontogenetically has migrated far toward the rear (Schultz 1942a, 1955). All these conditions are as yet barely indicated in older fetuses of these apes. For the attachment of the dorsal neck muscles the spinal processes of the cervical vertebrae gradually acquire an extreme length (Figures 17 and 18) and the occipital plane is tilted from a nearly level direction in the newborn ape (Figure 19) to a perpendicular one in the adult, in which it often becomes even enlarged at the top by a transverse bony crest.

CRANIAL GROWTH

Figures 20 and 21 illustrate corresponding cranial age changes in hylobatids, particularly the tilting of the nuchal part of the occiput and the shifting of the occipital condyles toward the back, besides the forward

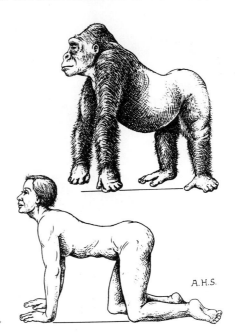

FIGURE 15
Sketches of an adult gorilla and an adult woman with trunk held about horizontally and looking straight forward, showing the many differences in proportions and in the position of the shoulder and formation of the neck.

A.H.S.

Orang-utan

A.H.S.

FIGURE 16
Adult female orangutan carrying its offspring (after Schultz 1969a).

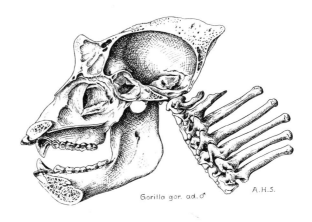

FIGURE 17
Midsagittal section of the skull of a wild adult male gorilla and the cervical vertebrae.

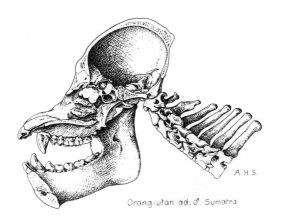

FIGURE 18
Midsagittal section of the skull of a wild adult male orangutan and the cervical vertebrae.

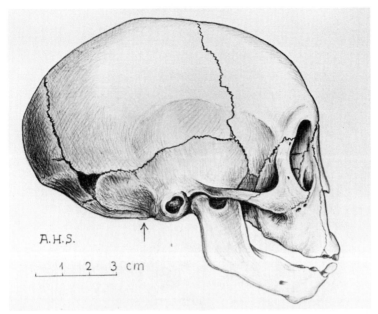

FIGURE 19
Skull of a newborn wild gorilla. The arrow points to the center of the occipital condyle.

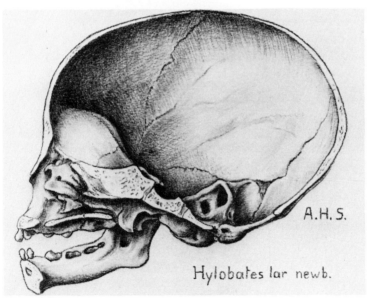

FIGURE 20
Midsagittal section of the skull of a newborn gibbon (after Schultz 1973a).

271

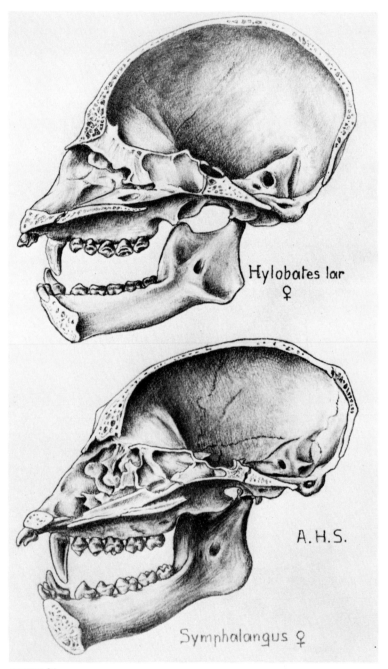

FIGURE 21
Midsagittal sections of skulls of adult wild gibbon and siamang (after Schultz 1973a).

migration of the face. It is of special interest that a central position of the craniovertebral joint is characteristic of all simian primates during early stages of their development and that subsequently precondylar growth of the skull far surpasses postcondylar growth, except in man. In prosimians, however, the position of this joint, and hence of the foramen magnum, undergoes no noteworthy ontogenetic change, always being near or at the rear of the skull, as in the smallest and largest prosimians, shown in Figure 22. The superimposed sections of skulls of infantile, juvenile, and adult chimpanzees in Figure 23 demonstrate most convincingly these typical simian age changes in the relative position of the occipital condyles and in that of the entire splanchnocranium which gradually moves forward from underneath the braincase, the latter process being most pronounced in males. In chimpanzees this sex difference becomes nowhere near as great as it does in the adult mandrills illustrated in Figure 24. The sectioned skull of an older male in Figure 25 shows even more clearly the extreme postnatal forward displacement of the powerful jaws, supported by bilateral bony nasal ridges and a remarkably thick mandible. In mandrills the males reach twice the weight of females during late postnatal growth. The same occurs in proboscis monkeys in which the sexes become further differentiated by

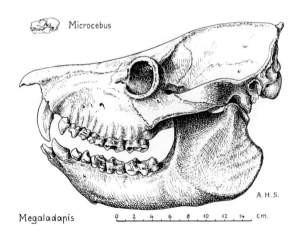

FIGURE 22
Skulls of an adult *Microcebus* and an adult *Megaladapis,* drawn to the same scale (after Schultz, 1969a).

273

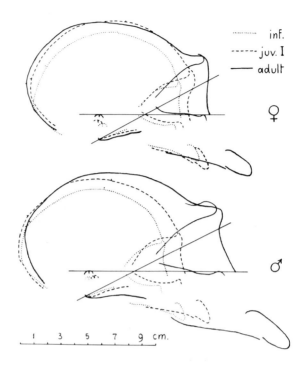

FIGURE 23
Superimposed midsagittal sections and sections of orbits of
chimpanzee skulls of female and male infants, juveniles, and
adults, oriented according to nasion—basion line with basion
coinciding (after Schultz 1969b).

the relative size of the outer nose, where the intensely vascularized wing
portion of males changes from its small fetal beginning into the grotesque
organ illustrated in Figure 26 (Schultz 1942b). It is a well-founded conclusion that such sexual dimorphism, due to different rates or duration of
postnatal growth, is lacking or minimal in prosimians as well as nearly all
platyrrhines, whereas it is maximal among a great variety of Old World
monkeys and apes (Schultz 1962, 1970). This is most easily apparent in
regard to general body size and the relative length of the permanent canines
in most catarrhines.

TEETH AND SKULLS

The average total length of the completed postcanine dentition differs
surprisingly little according to sex (Schultz 1962, 1973a). That the growth
of the dentition and that of the jaws are not as closely correlated as might be

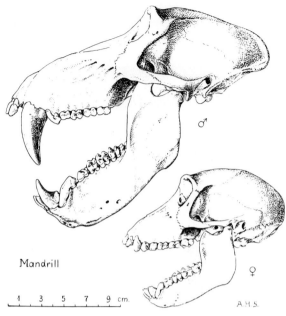

FIGURE 24
Skulls of an adult male and an adult female mandrill, drawn to the same scale (after Schultz 1969a).

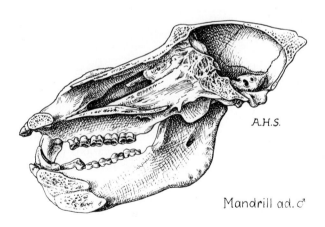

FIGURE 25
Midsagittal section of the skull of an old wild mandrill.

275

Nasalis

A.H.S.

FIGURE 26
Head of a wild adult male proboscis monkey (after Schultz 1969a).

expected, is shown by the comparison between the two male West African gorillas in Figure 27. One is extremely macrodont and the other almost microdont, yet the entire palate of the latter is much longer than that of the former. Further examples of such marked differences in the size of the teeth and their supporting bony structures within the same primate species have been illustrated by the writer in other publications (1963, 1969). The remarkable independence of dental and bony growth has also been convincingly demonstrated by Tonge and McCance (1973) in pigs by means of experimental undernutrition which leads to unequal growth of the dentition and the jaws respectively, thereby producing overcrowding and even suppression of teeth. Crowded and otherwise malformed dentitions are not at all rare among wild primates and have been found by the author (1935, 1972) in a great many specimens of which a few are shown in Figure 28. Of special interest are cases of reduction and total absence of teeth that are

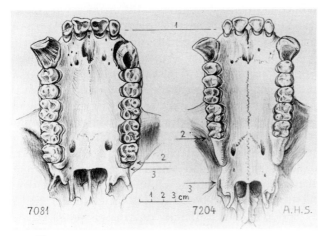

FIGURE 27
Upper jaws of adult male West African gorillas with marked differences in
size of teeth and palates, though the basion—prosthion distance happens
to be the same (205 mm.) in both.

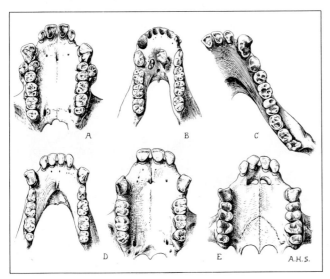

FIGURE 28
Examples of abnormal development in the permanent dentition of wild
primates: A = duplication of both first premolars and reduced third
molars of a male orang-utan; B = male gorilla with displaced lower
premolars; C = male orang-utan with supernumerary molars (5 on r.,
4 on l.); D = female chimpanzee with congenital lack of two lower
and all upper premolars and of one upper incisor and with reduced
lower left third molar; E = female marmoset with congenital lack of
one premolar on both sides (after Schultz 1956b).

typical for some major groups of primates while having become normally eliminated in other groups. For instance, the marmoset in Figure 28 E has only two premolars, as is normal for catarrhines, and so has the spider monkey in Figure 29 on one side of the lower jaw. The capuchin monkey in Figure 29, as well as the spider monkey, show congenital lack of third molars as in the Callithricidae. While such occurrences can gradually become more frequent in some populations, the varied other conditions of abnormal development in the dental apparatus have evidently never been subjected to favoring selection. Many other phylogenetic innovations in the skulls of primates can be regarded as results of ontogenetic alterations in the relative rates of growth of certain cranial parts. For instance, in the great majority of human skulls the alisphenoid and parietal bones approach each other during their early ossification and thereby prevent the temporal and frontal bones from forming contact during the closure of the lateral fontanelles, as they do in most nonhuman catarrhines. Similarly, the lacrimal and ethmoid are wedged apart by extensions of the maxillary and frontal bones in many chimpanzees and gorillas as well as in exceptional cases of man, but never in any other simian primates (Schultz 1950b, 1968). As indicated by the examples in Figure 30, the topographic relations, which form the rule for man, are also typical for *Pan paniscus,* whereas very rare in *Pan troglodytes* (Schultz 1969b, Vandenbroek 1969), showing that such specific differences can develop readily within local populations through slight changes in the speed and direction of early ossification. It may also be

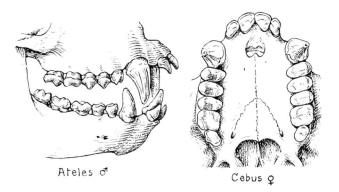

Ateles ♂

Cebus ♀

FIGURE 29
Wild adult *Ateles geoffroyi* with congenital lack of both upper third molars and of lower right first premolar and with lower right canine occluding behind upper canine. Wild adult *Cebus albifrons* with congenital lack of all third molars and upper second molars much reduced (after Schultz 1972).

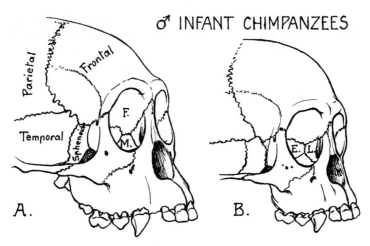

FIGURE 30
Skulls of infantile chimpanzees with different relations between the bones in the temple and in the medial orbital wall. A = Cameroon, B = South of Congo river. E = ethmoid, F = frontal, L = lacrimal, M = maxillary (after Schultz 1950b).

mentioned in this connection that large mastoid processes, supposedly distinguishing recent man, actually become just as large in old African apes, but only after they had reached adulthood in sharp contrast to the early appearance of these processes during infantile life of man (Schultz 1952). The relative ages of fusion between the centrale and naviculare in the wrist and of the coalescence between the intercostal segments of the sternum, differ widely among simian primates and even the hominoids and hence, represent further examples of ontogenetic distinctions responsible for the varying grades of specialization seen in adults (Schultz 1936, 1956a, 1961).

That the tempo or duration of growth in specific bodily parts and the relative age at which developmental processes occur are closely connected with evolutionary changes has been demonstrated for innumerable other characters. The promising significance of all these findings has not yet received adequate attention, probably because we still know very little of the real causes of changes in ontogenetic schedules and merely can point out their results. It seems obvious, for instance, that the discovery of the *reason* for the restricted, but very great, difference in the prenatal growth rate of the caudal region of stumptailed and of longtailed macaques would be more satisfying than merely to label such a distinction between closely allied species as *adaptations* to some diverging modes of life. As another example it might be mentioned that we can only demonstrate the phylogenetic increase in brain size of adult hominids with the many recent attempts to estimate the capacity of fossil crania, but the basic causes of this quite rapid

279

change might ultimately be found in many more comparative data for the postnatal growth of the cranial cavity in recent hominoids, which so far have indicated a decided human distinction (Schultz 1965). Unfortunately, the systematic collection of information on the intraspecifically variable age changes from early fetal to adult life is still much hampered by the scarcity of suitable material in adequate numbers and proper preservation.

In conclusion, the writer hopes that these brief notes, accompanied by a random selection of his drawings, may at least have shown the justification and need for saving specimens at all stages of their development, and that they may also induce younger investigators to undertake new and undoubtedly rewarding studies of primate ontogeny in its bearing on phylogeny.

Editor's note: The following four photographs (Figures 31–34) show Dr. Schultz at work.

FIGURE 31
Dr. Schultz weighing an adult gibbon. Thailand 1937.

FIGURE 32
Dr. Schultz examining an infant gibbon.
Thailand 1937.

FIGURE 33
Dr. Schultz holding a giant hornbill. Thailand 1937.

FIGURE 34
Dr. Schultz making a facial cast of an adult
orangutan. An interesting note is that the furniture
in the laboratory shown here was made by Dr.
Schultz. Borneo 1937.

References Cited

Beer, G. R. De
1930   Embryology and Evolution. Oxford: Clarendon Press.

Bolk, L.
1926   Das Problem der Menschwerdung. Jena: Fischer.

Keibel, F.
1906   Die äussere Körperform und der Entwickelungsgrad der Organe bei Affenembryonen. Selenka: Menschenaffen-Studien über Entwickelung und Schädelbau. 9. Liefer. Wiesbaden: Kreidel.

Leutenegger, W.
1973   Maternal-fetal weight relationships in primates. Folia Primatologia 20.

Schultz, A. H.
1924   Observations on Colobus fetuses. Bulletin American Museum of Natural History 49:443–457.

1926   Fetal growth of man and other primates. Quarterly Review of Biology 1:465–521.

1933   Die Körperproportionen der erwachsenen catarrhinen Primaten mit spezieller Berücksichtigung der Menschenaffen. Anthrop. Anz. 10:154–185.

1935   Eruption and decay of the permanent teeth in primates. American Journal of Physical Anthropology 19:489–581.

1936   Characters common to higher primates and characters specific for man. Quarterly Review of Biology 11:259–283, 425–455.

1941   Chevron bones in adult man. American Journal of Physical Anthropology 28:91–97.

1942a   Conditions for balancing the head in primates. American Journal of Physical Anthropology 29:484–497.

1942b   Growth and development of the proboscis monkey. Bulletin Museum of Comparative Zoology (Harvard University) 89:279–314.

1949   Sex differences in the pelves of primates. American Journal of Physical Anthropology n.s. 7:401–424.

1950a   The physical distinctions of man. Proceedings, American Philosophical Society 94:428–449.

1950b   The specializations of man and his place among the catarrhine primates. Cold Spring Harbor Symposia on Quantitative Biology 15:37–53.

1952   Über das Wachstum der Warzenfortsätze beim Menschen und den Menschenaffen, mit kürzer Zusammenfassung anderer ontogenetischer Spezialisationen der Primaten. Homo 3:105–109.

1954   Studien über die Wirbelzahlen und die Körperproportionen von Halbaffen. Vierteljahrsschr. Naturforsch. Ges. Zurich 99:39–75.

1955   The position of the occipital condyles and of the face relative to the skull base in primates. American Journal of Physical Anthropology n.s. 13:97:120.

1956a Postembryonic age changes. Primatologia 1:887–964. Basel: Karger.

1956b The occurrence and frequency of pathological and teratological conditions and of twinning among non-human primates. Primatologia 1:965–1014. Basel: Karger.

1961 Vertebral column and thorax. Primatologia 4, Liefer. 5:1–66. Basel: Karger.

1962 Metric age changes and sex differences in primate skulls. Z. Morphol. u. Anthrop. 52:239–255.

1963 Age changes, sex differences and variability as factors in the classification of primates. *In* Classification and Human Evolution. S. L. Washburn, ed. New York: Wenner-Gren. Viking Fund Publications in Anthropology 37:85–115.

1965 The cranial capacity and the orbital volume of hominoids according to age and sex. *In* Homenaje a Juan Comas en su 65 Aniversario 2:337–357. Mexico.

1968 The recent hominoid primates. *In* Perspectives on Human Evolution. Vol. 1, pp. 122–195. New York: Holt, Rinehart and Winston.

1969a The Life of Primates. London: Weidenfeld and Nicolson.

1969b The skeleton of the chimpanzee. *In* The Chimpanzee. Vol. 1, pp. 50–103. Basel: Karger.

1970 The comparative uniformity of the Cercopithecoidea. *In* Old World Monkeys. J. R. Napier and P. H. Napier, eds. Pp. 39–51. London: Academic Press.

1972 Developmental abnormalities. *In* Pathology of Simian Primates. Vol. 2, pp. 158–189. Basel: Karger.

1973a The skeleton of the Hylobatidae and other observations on their morphology. *In* Gibbon and Siamang. Vol. 2, pp. 1–54. Basel: Karger.

1973b Age changes, variability and generic differences in body proportions of recent hominoids. Folia Primatologia 19:338–359.

Starck, D.
1962 Der heutige Stand des Fetalisationsproblems. Hamburg: Paul Parey.

Tonge, C. H. and R. A. McCance.
1973 Normal development of the jaws and teeth in pigs, and the delay and malocclusion produced by caloric deficiencies. J. Anat. 115:1–22.

Vendenbroek, G.
1969 Evolution des Vertébrés. Paris: Masson & Cie.

Dorothy Lamour and Jiggs, stars of *Jungle Love,* relax between scenes. © 1938 National Geographic Society.

S. L. Washburn and
Elizabeth R. McCown

Human Evolution and Social Science

This paper offers three examples showing the importance of a biosocial approach to the study of behavior. No social system can be fully understood without an understanding of the actors in that system, whether they are termites or human beings. The addition of human culture complicates the picture but does not remove the biological element. Human biosocial adaptation includes the errors inevitably made by the primitive brain, and these errors are built into the foundations of every culture.

BACKGROUND

Some years ago it appeared that social anthropology would develop into a powerful system of scientific thought. A society for applied anthropology was founded. The American Anthropological Association was reorganized. The social sciences in general underwent an enormous expansion. Yet no

This paper was delivered at the annual meeting of the American Association of Physical Anthropologists in St. Louis, April 15, 1976. We wish to thank Charles M. Nelson for the grant that supported this work.

285

great change in social science occurred. During a period of remarkable advances in astronomy, physics, and biology, the social sciences have undergone no revolution. Recent textbooks contain little that could not have been easily understood 40 years ago.

There may be something wrong in the very nature of the social sciences. At present, theorists in this field cannot develop a powerful system of scientific thought. Suppose that the next innovations in social science are not going to come from social science at all. Possibly, foundations of the behavioral science of the future will be biological. There are strong trends in this direction—in psychiatry, animal behavior, and the embattled science of sociobiology.

An indication of the newer interests and directions was shown in the sessions at the meetings of the American Anthropological Association in San Francisco in November 1975. There were fewer papers on physical anthropology—using the term in the traditional sense—and more with a biological cast than ever before. There were strong biological interests in sessions on ethology, primatology, medical anthropology, aggression, and biogenetic structuralism. Both medical anthropology and biogenetic structuralism occupied entire days. At least three biobehavioral symposia were presented at the American Anthropological Association meetings in Washington D.C. in November 1976.

A series of new books has appeared in the last two years: *Animal Behavior* (Alcock 1975); *Evolution of Behavior* (Brown 1975); *Biosocial Anthropology* (Fox 1975); *Animal Nature and Human Nature* (Thorpe 1974); *From Instinct to Identity* (Breger 1974); *Biogenetic Structuralism* (Laughlin and D'Aquili 1974); and *Sociobiology* (Wilson 1975).

It is our belief that the climate of opinion has really changed, and that evolution (particularly human evolution) and animal behavior (particularly primate behavior) are now popular and influential. The change is probably due to television, numerous successful animal programs, and popular books. But this success parallels a rapidly rising scientific interest in behavioral evolution as a way of looking at the nature of the world and of man.

These changes offer great opportunities to physical anthropology. In the bibliographies of the new books cited above, there are many references to the contributions of anthropologists. The data on human evolution is of great interest and is being widely used by many scientists in marked contrast to the situation a few years ago.

Writers such as Alexander, Bigelow, or Wilson tend to jump from genetics and the behaviors of contemporary forms directly to conclusions on human evolution. Oddly, they are minimally concerned with the fossil record, archeology, adaptation to climate, or anatomical evidence on the behavior of earlier forms of man.

In a sense, physical anthropology fits between the theories and the conclusions. It provides the facts about human evolution, and the more important sociobiology becomes, the more the data on the fossil record will be needed. But with the greatly renewed interest in human evolution and animal behavior comes a set of problems—most clearly seen in the controversies surrounding Wilson's *Sociobiology* (Wade 1976), but by no means limited to them. The biosocial problems of: social living, intelligence, sex, aggression, to name only a few, are profoundly controversial. At a philosophical level, these cut to the heart of the nature of man and of the process that produced him. At the personal level, these are inherent in the problems of daily life, and inevitably involve ethics and the nature of social systems.

At a methodological level, these and similar problems are all subject to experimental analysis, and it is apparent that in the sociobiology of the future, unsupported opinion will be far less important than it is today. Most social scientists shrink from the very notion of experiments, and social scientists often repudiate the goal of experimental science. No short paper can do more than introduce some of the issues. This paper is introduced with a dogmatic statement on the interpretation of animal behavior, and hopefully, it will clarify and illustrate that statement with remarks on human behavior.

CONCLUSIONS

First, the analysis of any system of animal behavior must include an understanding of the biology of the actors in that system. Biology is much more complex than social systems. For example, primate social systems are far less complicated than primate brains. Second, human beings are no exception to this rule. Third, the reason that the social sciences cannot become scientific—in the sense that biology is—is because the fundamental biology of human beings is excluded.

In this paper three examples will be used to show the importance of a biosocial approach to the problems of human behavior. Hopefully, the same three examples will show why social science, as it is practiced today, cannot be scientific in the same sense that biology is.

The theme will be that the human biology, particularly the brain evolved because of success in adapting to the prescientific world. The result of the biosocial adaptation to the primitive world is that human beings make fundamental mistakes; they misunderstand time, space, cause, and the nature of social systems. These misunderstandings are built into the foundations of every culture.

The Primitive World versus the Modern World

What the sciences do, considering biological science particularly, is attempt to remove the mistaken, inaccurate common-sense notions. Astronomy devised methods to demonstrate that the earth is not flat and is not the center of the universe. Chemistry developed methods proving that the elements are not earth, air, fire, and water. Biology showed that, contrary to common sense and philosophy, introspection is an extraordinarily limited intellectual tool, unable even to analyze the words with which we communicate.

But social scientists are content to describe and theorize. There is no effort to remove the errors of the past. For example, it is natural for human beings to think that there are spirits, and such beliefs are organized in an extraordinary variety of ways. It is the business of ethnography to describe these customs, and the social scientist thinks about the meanings of the symbolic systems. But the point is that there are no such spirits. They form a part of the belief system of the primitive world.

Just as the unaided human brain naturally believes that the small, flat world is made for the convenience of man, so, in a similar way, the brain peoples the primitive world with deities and spirits. In biology it is thought that the business of the scientist is to escape from the bondage of the primitive world. But in social science the aim is to describe, to honor cultural relativity, and to hide behind the theory that cultures are complex. Social science is a part of the primitive world, and the same customs that could not control the bow and arrow are the subject matter of social science in the nuclear age. As we see it, the fundamental reason that it is difficult for biologists and social scientists to communicate effectively is because one group wants to get out of the primitive world and the other wants to stay in it.

For example, spontaneous generation is a natural belief. Prior to modern biological science, probably all peoples believed in spontaneous generation. A recent note in *Current Anthropology* (Fuchs 1973) states that the world view of the Australian aboriginals is unrealistic. The world is not full of spirits and ghosts, nor do women need the intervention of spirits to become pregnant. Fertility cults are not indispensable for rain. Black magic and witchcraft are not really effective. The aboriginals built for themselves a world of illusion. Much the same could be said for all peoples until very recently, and it is this fantasy world that prevents social science from being scientific.

The fantasies of social life are what the human brain does with social facts, before there is scientific knowledge. The brain is both rational and irrational at the same time. Given the conditions of human life before agriculture, the brain made successful adaptations. But, judged by what is now known, it also made fundamental mistakes. Adaptation is limited by knowledge. The brain adapts to a humanly perceived world, but before modern science the perception of the world had major limitations. As MacLean has clearly stated, the brain is both logical and emotional. The aborigines had useful knowledge, but to this they added the fantasies of emotional cerebration. MacLean has pointed out that it is this emotional cerebration that has the characteristic of being able to stand on both sides of any question and to give the feeling of "being real, true and important" (MacLean 1970).

From an evolutionary point of view we can add to this statement the explanation that the brain must package information in this way, because throughout human history people had to act. Acts involve emotion. Human societies could not wait for accurate information, but had to act on the basis of the apparent facts of the primitive world.

Once these apparent facts are packaged by the human brain they then become "real, true, and important." They are accepted and may no longer be doubted, and people who hold other beliefs are regarded as barbarians, perhaps hardly human.

If, following Durkheim, the religious system is viewed as an expression of the social system, then there necessarily were many religious systems. There were hundreds of social systems, based in part on fantasy, and each one was regarded as real. Religion is basically a system of traditional power (Bloch 1974), and it is what makes the political system seem an undeniable fact. Our culturally determined view that religion is primarily a system of explanation is wrong, and the "awe" of much religion is the very real power of the elders. Today the problem for social science is that all cultures, including our own, are products of the prescientific world. All cultures are built on fantasy as well as fact. And in all cultures, power is used to enforce the local fantasy.

The civilization of the Aztecs was a good example. They had a great imperial court, they built cities, markets, gardens. The mixture of religion with power was never clearer. But the great pyramids were built so that thousands of human beings could be sacrificed to nonexistent gods. It took nearly constant warfare to provide the number—some fifteen thousand people—who were needed for sacrifice each year. We can look with wonder on the achievements of the Aztec empire, or we can see in their system the evidence of misconceptions, and the lengths to which the human brain may easily go to rationalize a mad, mad world. The Aztec civilization shows why the unaided human brain is unreliable. From the point of view of the individual in the system, the human brain can *make almost any system seem natural.* From the point of view of the outsider looking at the system, any system may be analyzed in several different ways.

In looking at a social system, the outside scholar may emphasize evolution, history, material factors, human nature, or language and symbols. The fashions of interpretation may have no more validity than the beliefs of the natives. One point should be made clear: the study of all peoples is interesting and important. Ethnology, the study of the diversity of human customs, is a fascinating subject matter; and comprehensive description is necessary. Biological explanations are frequently regarded as reductionist. Actually, biological and social explanations supplement each other. For example, bipedal walking is a biological fact that is basic to many human behaviors and that differentiates man from ape. Understanding the biology of walking gives no indication of all the possible cultural elaborations in hopping, skipping, dancing, gymnastics, and as essential parts of numerous games. Biology calls attention to the fact that the fastest human runners run less than half as fast as the social carnivores, that human hunting cannot be carried out in ways similar to that of the carnivores, and that to hunt like a wild dog, a human female would have to run a two-minute mile carrying a baby.

But we started speaking of a powerful social science, of applied social science. Looking to the future in a time of rapid change, what social science can be relied on? Even economics, the most quantitative and scientific of the social sciences, has usually been wrong in its predictions. Lekachman's 1976 *Economists at Bay* suggests that this is because the basic economic problems need to be stated in very different form.

Medical Anthropology

Perhaps the issues may be made clear by considering medical anthropology. Disease has always been a major problem for human beings. Every culture has theories of diseases and cures; basically the theories are very similar. The diseases are perceived to be caused by spirits, magic, or some person. This is simply a special case of the way the brain manages causes. Natural causes are seen in personal terms. This is the human application of a very general situation in animal behavior. Animals act *as if* other animals, or aspects of nature, were the same as they are. When dogs and monkeys interact, each species continues its own behavior. There were hundreds of tribes, and scholars could spend many, many lifetimes gathering information on all the different beliefs on the cause and cure of disease. Part of medical anthropology is concerned with this kind of information.

Another part of medical anthropology is concerned with the delivery of medical information and services. All the traditional customs will not cure disease. With very few exceptions, cures are the products of modern science, and the understanding of medical practices is helped by some biological knowledge. Anthropologists are divided on whether they want to devote their efforts to collecting medical folklore, or helping people to be healthy. Social science only has a role in the second—in helping—and the endless collection of medical customs is more likely to impede the spread of medical science than to help it. The fact that beliefs are universal does not mean that they are correct. Prescientific beliefs about disease are remarkably similar but are uniformly wrong. The common elements responsible for these beliefs are human biology and ignorance, and this combination will produce a wide variety of cultural universals.

The difference between biology and anthropology is clearly shown in the disputes over Lysenko. It seemed clear to biologists that what he said was wrong and it should be opposed. Knowledge gave a social obligation. Anthropologists feel no obligation to help people to understand where customs are wrong. As anthropologists (and other social scientists) look to the future, they have a choice (as much as any of us have choices) as to whether they want to be a part of the scientific revolution that has replaced the primitive notions of time, space, nature, and man, or to continue to describe and theorize. To make progress it is necessary to add biologic understanding to that of the social sciences. The present dilemma can be illustrated by considering the horror that was the Aztecs. Most of human history seems, in retrospect, to be almost equally uninviting. Human evolution did not produce cultures that are efficient, or, in any modern sense, necessary or pleasant.

We have tried to illustrate the influence of modern science through the case of medicine. It is technical knowledge that cures diseases, not ancient custom. We believe that a combined biosocial approach may be the most powerful method of moving toward the solution of many problems in the social sciences. To illustrate this point we have chosen the problem of understanding language.

Language

As described by Lancaster, communication in the nonhuman primates is limited. Gestures are more important than sounds, and most messages are based on a combination of gesture and sound. What is conveyed is primarily the emotional state of the sender. Experimentally, the gestural part of the system may be elaborated under human guidance. So far, at least, no one has succeeded in teaching any nonhuman primate to speak.

Recent experiments by Ploog (1970), Robinson (in press), and particularly Myers (Chapter 3), have shown that sounds in nonhuman primates are controlled primarily by the limbic system, and that the cortex is unimportant in their emotional communication.

In monkeys, massive removal of cortex does not affect vocalization, and only slightly affects facial expression. Electrodes implanted in the limbic system may elicit all the normal sounds.

In man, the sounds of language are controlled by the cortex, and meaning is conveyed by a phonetic code. It is this code that makes the system open, allowing an almost infinite number of combinations. No nonhuman primate uses a code, and, so far, no one has succeeded in teaching one to an ape.

In briefest outline then, the origin of language is the evolution of the ability of the brain to make and understand complex patterns of sounds composed of phonemes. Because of his biology, man easily learns language, and no other animal has a comparable adaptation. Even a mynah bird, with its great ability to copy sounds, does not use a sound code.

Understanding the origin of language is a biosocial problem. All the descriptive efforts of linguists could not even formulate the problem because the brain was not considered. But the facilitating biology gives no understanding of the diversity and complexity of languages—the subject matter of linguistics. But it is the biology that makes possible the learning of extraordinarily different linguistic systems. It does not determine what is learned in any detailed way. The learning is limited, and, probably, much of what is referred to as the *deep structures* of language cannot be interpreted without biologic understanding.

But the relations between biology and language are much more subtle than just the biology making language learning possible. People talk, but they do not understand that they are using a phonetic code. Various forms of picture writing were used for more than two thousand years before the alphabet was invented. Even today, in China and Japan, nonphonetic writing may be used. Many millions of man hours have been spent unnecessarily because people did not understand the phonetic basis of the language they were using. The absurdities of English spelling are a time-wasting relic of linguistic ignorance. Because we use a learned social tradition (in this case, language) and feel strongly about it (the emotional brain) does not mean that we understand the custom. People have the illusion that they understand their own customs, but this is only true in a superficial way. The ability to speak gives no understanding of the facilitating mechanism.

In monkeys or human beings with divided cerebral hemispheres, both sides of the brain think. Both can solve problems. But human language is controlled by one side, usually the left. There is a widely held belief that lnaguage is necessary for thought, but the divided brain experiments have clearly shown that this is not the case.

Language is learned, but the learning is only possible because of the evolution of large and specialized parts of the brain. The biology of the behavior cannot be understood by the description of the behavior, no matter how carefully this is done. Brains are organized in very different ways, and, for the purposes of sociobiology, the organization must be understood (contrary to the beliefs of behaviorism). Brains are not just bigger or smaller; they are organized in very different ways.

Biology shows that generalization is very different from the way it is viewed in our culture, particularly in the descriptions of how scientists work. The brain generalizes from remarkably few facts, and then the generalizations are revised in the light of subsequent experiences. The more primitive the brain, the more it generalizes its sensory world. The more primitive the brain, the fewer the categories, and the less modification in the light of experiences. An animal could not accumulate and store large numbers of experiences and then make generalizations. It would be dead long before the information was generalized.

The biology of language is not just something remote, useful in understanding the evolution of communication. The biology of language is needed every day in the understanding of: communication, thought, language and thought, rational and emotional cerebration, logic, spelling, aphasias, and the verbal part of intelligence.

Toward a More Effective Social Science

At every point biology suggests that our culture's view of these matters, our common sense, is wrong. This is the challenge of biosocial anthropology, of sociobiology. The hope of sociobiology is to bring a different order of understanding to the study of human behavior: in part, through experimentation and the development of methods to supplement, to guide the function of the human brain; in part, by the elimination of the fantasies that all have lived by, by the elimination of the beliefs and customs of the primitive world; in part, by the creation of a new humanistic biology that recognizes the importance of both the individual and the social system.

Through their interests in history, customs, individuals, and biology, anthropologists have a unique opportunity to create a far more effective science of man. The essential point is simply that the process of evolution through natural selection produces adaptation. It produces competitive success. But it does not necessarily produce ideal solutions, efficient systems, or anything beyond adaptation. The results of adaptation, just what survived, are the myriad languages and cultures of the primitive world. The modern relics of that world are warfare, poverty, disease, and social futility.

The world of science is built on almost completely different foundations. There are universal understandings in the sciences. These understandings create health, wealth, and the possibility of efficient social systems. Perhaps this new world may be used wisely for the benefit of all mankind. To accomplish this, the beliefs and foundations of the primitive world must be removed and they must be replaced by new, experimentally verified knowledge, just as has been done in the other sciences.

At the present moment, a clarification of the issues and a step toward a more effective social science is most likely to come from the study of animal behavior, from the study of human evolution, and from the biological nature of man.

References Cited

Alcock, John
1975   Animal Behavior. Sunderland, Massachusetts: Sinauer Associates.

Alexander, Richard D.
1974   The evolution of social behavior. Annual Review of Ecology and Systematics 5:325–383.

Bigelow, Robert
1969   The Dawn Warriors: Man's Evolution toward Peace. Boston and Toronto: Little, Brown.

Bloch, Maurice
1974   Symbols, song, dance and features of articulation: Is religion an extreme form of traditional authority? European Journal of Sociology 15:55–81.

Breger, Louis
1974   From Instinct to Identity: The Development of Personality. Englewood Cliffs, New Jersey: Prentice-Hall.

Brown, Jerram L.
1975   The Evolution of Behavior. New York: W. W. Norton.

Fox, Robin, ed.
1975   Biosocial Anthropology. London: Malaby Press.

Lancaster, Jane B.
1975   Primate Behavior and the Emergence of Human Culture. Basic Anthropology Unit Series. George and Louise Spindler, eds. New York: Holt, Rinehart and Winston.

Laughlin Jr., Charles D. and Eugene G. d'Aquili
1974   Biogenetic Structuralism. New York and London: Columbia University Press.

Lekachman, Robert
1976   Economists at Bay. New York: McGraw-Hill.

MacLean, Paul D.
1970   The triune brain, emotion, and scientific bias. In The Neurosciences: Second Study Program, F. O. Schmitt, ed. Pp. 336–349. New York: Rockefeller University Press.

Ploog, Detlev
    1970   Social communication among animals. *In* The Neurosciences: Second
            Study Program. F. O. Schmitt, ed. Pp. 349–361. New York: Rocke-
            feller University Press.

Robinson, Bryan W.
    In   Limbic influences on human speech. Paper presented at Conference on
    press Origins and Evolution of Language and Speech, New York Academy of
            Sciences, September 1975.

Thorpe, W. H.
    1974   Animal Nature and Human Nature. Garden City, New York: Anchor
            Press/Doubleday.

Wade, Nicholas
    1976   Sociobiology: troubled birth for new discipline. Science 191:1151–1153.
            1153.

Wilson, Edward O.
    1975   Sociobiology: The New Synthesis. Cambridge, Massachusetts: Harvard
            University Press.

The Authors

Frank A. Beach, after receiving his Ph.D. from the University of Chicago in 1940, spent ten years at the American Museum of Natural History as Curator of Animal Behavior. His research on brain and instinctive behavior led to his appointment as Sterling Professor of Psychology at Yale University. In 1958 he moved to the University of California at Berkeley where he continued his research on hormonal control of sexual and parental behavior in animals. His comparative studies range over a wide variety of mammals, bird, and reptiles, including the evolutionary origins of human sexuality.

Mary LeCron Foster, a linguist trained at the University of California at Berkeley, is currently a research associate in the Department of Anthropology at the University of California. Her earliest linguistic interest was in Mexico, where she has done fieldwork annually since 1958. For the last 13 years she has delved into the past in her efforts to reconstruct man's primordial language. While her conclusions are bound to be controversial, they are of great significance in helping to unravel the role of language in the evolutionary process leading to modern man.

Beatrix A. Hamburg took leave from Stanford University in May 1976 to assist at the National Institute of Mental Health in the development and monitoring of research programs in adolescence and childhood. Her major research interests and publications are in the areas of coping with stress; biocultural aspects of sex role development;

hormonal and behavioral changes in adolescence; early adolescence as a stressful and distinctive phase of the life cycle, and preventative intervention in the schools to teach coping skills to adolescents in a peer counseling program.

Prior to this Dr. Hamburg was Associate Professor in the Department of Psychiatry and Behavioral Sciences at Stanford. There, since 1961, she divided her time among clinical, administrative, teaching and research responsibilities in child psychiatry. In 1973, she resigned as Director of the Child Psychiatry Clinic to devote more time to research and to become Associate Director of the Laboratory of Stress and Conflict.

Wilton Marion Krogman's first love in Anthropology was in Archaeology, with a special interest in exhumed skeletal remains, first of the American Indian (Illinois), then of the proto- and early historic peoples of the Middle East (Anatolia, Palestine, Jordan, Iran). In the 1940s he turned to research in the physical growth of the American child, and in 1947 he founded the Philadelphia Center for Research in Child Growth, (later renamed the Krogman Growth Center). In 1971 he was retired from the University of Pennsylvania as Emeritus Professor of Physical Anthropology, School of Medicine. Since then he has been Director of Research, Lancaster (PA) Cleft Palate Clinic. Throughout his life he has devoted much of his spare time to Forensic Anthropology, specializing in the identification of human skeletal remains.

Elizabeth R. McCown is a physical anthropologist associated with the University of California, Berkeley. Her principal interest is comparative anatomy of primates. Since 1973 she has been managing editor of the Perspectives on Human Evolution series.

Paul D. MacLean, a graduate of Yale College and Yale University School of Medicine, has been a research scientist since 1957 at the National Institute of Mental Health, where he is now Chief of the Laboratory of Brain Evolution and Behavior. In connection with his investigations, he has expounded the view that the human forebrain has evolved along the lines of three basic patterns that may be characterized as reptilian, paleomammalian, and neomammalian. Radically different in chemistry and structure, and in an evolutionary sense eons apart, the three formations constitute a *triune* brain with an amalgamation of three mentalities. Dr. MacLean is particularly interested in the epistemological implications of this situation, calling attention to intellectual perplexities that may arise because the reptilian and paleomammalian formations, which are involved in basic proclivities and emotions, lack the neural machinery for verbal communication.

Ernst Mayr, born and educated in Germany, came to the United States in 1931. He was first associated with the American Museum of Natural History, and was concerned with the collection and classification of birds, particularly of the Pacific area. His interests in natural history were combined with a remarkable understanding of genetics and evolutionary theory. He became one of the major figures in the new evolutionary synthesis of the 1930s and 1940s. As Agassiz Professor of Zoology at Harvard and later Director of the Museum of Comparative Zoology at Harvard, he retained his interest in classification, related it to the latest biology, and used it as a framework for understanding life. Author of numerous publications and recipient of many honors, he received the National Medal of Science in 1970, and is now Alexander Agassiz Professor Emeritus, Harvard University.

Ronald E. Myers is an M.D. and a neuroanatomist currently associated with the National Institutes of Health. He has been active in brain research since 1951. His early research demonstrated the role of corpus callosum in transmittal of visual learning and visual memories between the hemispheres. Prior to this research, no function could be clearly attributed to corpus callosum. His current interests focus on the brain pathologic effects of respiratory and circulatory disorders.

Adolph H. Schultz received his Ph.D. from the University of Zurich in 1916 and devoted his entire life to the study of the primates. From 1916 to 1925 Dr. Schultz worked at the Laboratory of Embryology of the Carnegie Institution, where he studied fetal growth in man and other primates. Until that time measurements were taken in a wide variety of ways, but, as a part of his investigations of fetal growth, Schultz standardized methods of measurement. He also undertook fieldwork in Nicaragua and later in Panama to build his collection of primate skeletons, since he had realized that relying solely on zoos and laboratories for specimens was a severe handicap.

In 1925 he left the Laboratory of Embryology and joined the Department of Anatomy at Johns Hopkins University where, with E. Huber, A. B. Howell, W. L. Straus, and G. B. Wislocki, he made fundamental contributions to the study of facial muscles, locomotion, reproduction, and numerous subjects in primate anatomy and human evolution. Schultz was among the first to understand the importance of following the growth of individual animals, and he kept a small colony of chimpanzees at Hopkins.

In 1936 Harold J. Coolidge organized an expedition to Southeast Asia to make collections for the Harvard Museum of Comparative Zoology. Schultz organized the collecting of gibbons, orangutans, and several species of monkeys, and personally supervised the preparation of specimens, measuring and making notes on every one. He illustrated all of his own papers, and these drawings are an essential part of his contribution to primatology. His rapidly done field notes are of extraordinary quality and form the basis of many published illustrations.

In 1933 Schultz was elected President of the American Association of Physical Anthropologists; he was elected to the National Academy of Sciences in 1939. In 1951 he returned to Zurich as Director of the Institute of Anthropology. He received the Viking Fund Medal in 1948 nd took part in numerous international meetings at the Wenner-Gren Foundation for Anthropological Research.

Schultz's interests in growth, proportions, and variability resulted in many publications, especially *The Life of Primates* (1969). This book is a splendid summary of a scientific lifetime, but it was by no means his final publication. His research continued nearly until his death in 1976, and his final contributions are still in press.

Sherwood L. Washburn is currently teaching at the University of California, Berkeley. His principal interest is the teaching of human evolution. In research he has developed experimental methods suitable for the analysis of the problems of human evolution. He has attempted to relate structure and function through the study of the evolution of behavioral systems. He became noted for his studies of human and ape anatomy.

R. V. S. Wright was born in England and after graduating from Cambridge University spent a brief period in Kenya before settling in Australia in 1962. His special interest is the archaeology and physical anthropology of the Australian region. He is at present Associate Professor in the Department of Anthropology, University of Sydney.

Index

Abang, 220–235
Abel, S., 183
ACTH, 185
activeness, sexual dimorphism in, 131
Adair, F. E., 183
adaptation, 25
    see also selection
adolescence, hormones and behavior in,
    161
    and sex differentiation, 127–129,
        134–137
adornment, 43–44
    see also display behavior
adrenalin, dimorphic secretion of, 185
adrenocortical hormone, 184
affect, *see* emotion
affectional bonds, 163
    see also pair-bonding, mother-child
    bond
affiliative behaviors, sexual dimorphism
    in, 193–197
Agassiz, L., 12, 13, 14, 17, 20, 22–24
aggressiveness, as a biosocial problem,
    286, 287

and display behavior, 45
elicitors of, 162
and evolution, 185–188
and facial expression, 63–64
female, 195
in hermaphrodites, 173–176
intersexual, 187–188
sexual dimorphism in, 131, 132,
    185–188
and social role, 140–141
agricultural societies, and adaptation, 160
    sex roles in, 164
    see also preagricultural society
Ainsworth, M. D., 193–195
Alexander, R. D., 286
Altmann, M., 178
altruistic behavior, 44
American Anthropological Association,
    285–286
analogy and language, 116, 118
analytical thinking, sexual dimorphism
    in, 133
anatomy, comparative, 16, 255–280
Andersen, H., 161

androgenital syndrome, 174
androgens, 126–127
    blocking of, 183–184
    sexual differentiation and, 170–171,
       173
    stress and production of, 185
    see also progestin, testosterone
anger, 51
Animal Behavior (Alcock), 286
Animal Nature and Human Nature (Thorp),
    286
animals, behavioral studies of, 38, 287,
    294
    experimentation with, 34, 41, 45
    vocalization and expression in,
       63–66, 68
    see also monkeys
anthropocentrism, 27
anthropology, 27
    applied, 285–286
antisocial behavior, sexual dimorphism in,
    131
anubis baboon, 149
ape-man concept, 240, 242
apes, pelvic development of, 263
    sexual dimorphism in, 274
aphasia, 61–62
appeasement signals, 162
approval-seeking behavior, 193
Arai, Y., 170
Arctocebus, 260–262
Aristotle, 21, 51
articulation and semantics, 78–79,
    110–114
Ateles geoffroyi, 278
attachment behavior, dimorphism in, 196
    vs. dependency, 196–197
    see also affectional bonds
attention-seeking behavior, 131, 193
Australopithecus, as toolmaker, 216–218,
    233–235
aye-aye, 262–263
Aztecs, 289, 291

baboon, 149
    menstrual synchrony in, 183
baculum, 145

Baker, S., 175–176, 186
Bakke, J. L., 183
Bandura, A., 187
Barcombe Mills skull, 249
Barzun, J., 25
Beach, F. A., 123–151, 171, 172, 173
Beer, G. R. De, 257
behavior, as adaptation, 155, 160
    affiliative, 193–197
    biosocial differences in humans, 159,
      287
    nonsexual, 173–174, 184–188
    prototypical forms of, 43–45
    reproductive, 128–137, 166–173,
      181–184
behavioral sciences, 286
behaviorism, 53
Benedek, T., 177–178
Bergson, H., 27
Berkeley, G., 38
Berlin, B., 82
Berndt, C. H., 161
Berndt, R. M., 161
Bigelow, R., 286
Biller, H. B., 158
binocular vision, 161
Biogenetic Structuralism (Laughlin and
    D'Aquili), 286
biosocial, 158–159
Biosocial Anthropology (Fox), 286
bipedalism, 290
    anatomy and, 268–269
    and human sexuality, 148–149
Bishop, N., 186, 187
Bittner, J., 171
blood pressure, sexual dimorphism in,
    127, 134
Blurton-Jones, N. G., 187
body hair, 162–163
body proportions, primate comparisons,
    263–268
Bolk, L., 257
Boole, G., 33
Boring, E. G., 52
Boucher de Perthes, J., 240
Bowlby, J., 193
brachiation and hand development,
    232–233

brain, centers of emotional expression, 60–61
    development of and hormones, 167, 169–173
    evolution of, 2, 4, 36, 161
    hemispheric dominance of, 188–193
    minor hemisphere abilities, 192–192
    ontogenesis of in *Homo sapiens*, 16, 263–265
    organization of and behavior, 293
    and regulation of non-sexual behavior, 173–174
    research questions, 50–53
    sexual dimorphism in, 126–127, 170–174, 188–193
    size of, 279–280
    and social activity, 288–289
    vocal mechanisms of, 66–72
breasts, female, 127, 146
breeding behavior, 44
Bremer, J., 184
Bridgewater Treatises, 14–15
Bridgman, P. W., 39
Bristol Zoo, 219–221
Broca, P., 45–46, 60–61
Broom, R., 217
Broverman, I. K., 165–166
Brown, W. L., 180
Bruce, H. M., 183
Buckland, W., 20
Buffery, A. W., 189
Buffon, G. L., 13, 26
Bull, H., 178
burial practices, 117–118

Cameron, J., 190
Canary Islands, 18, 25
capuchin monkey, 260, 278
Carpenter, C. R., 233
castration experiments, 184, 186
catarrhines, 258, 260, 263–264, 274, 298
catastrophism, 14, 18–20, 26–27
caudate nucleus, 42
Cebus albifrons, 278
cebus monkeys, 65

Center for Genetic Epistemology, 47
cerebral functions, asymmetry of, 188–193
cerebral hyperaemia, 133–134
Chambers, R., 13, 15, 16, 20
Chance, M., 34, 140, 143–144
Charles, E., 179
chevron bones, 257–258
childhood, 129
    learning and, 162
chimpanzee, anatomical studies of, 261, 264, 266, 273–274, 276
    communication experiments with, 71–72
    food sharing among, 163
    toolmaking among, 216
Chomsky, N., 82, 83
Christie, A. M., 191
chromosomes, 125
    and reproductive success, 166–167
    see also x chromosomes, y chromosomes
Ciochon, R. L., 52
Clemens, L. G., 172, 185
Clemetson, C. A., 178
clitoris, 126
    and female orgasm, 148–149
clothing manufacture, and sex roles, 164
cognate language forms, 90–110
Colobus, 260
coloration, 43–44
commissurotomy experiments, 189
communication, 39
    with chimpanzees, 71–72
    human vs. animal, 2–3
    and hunting, 161–162
    laws of, 38
    volitional, 62
    see also language, nonverbal behavior, prosematic behavior, speech, vocalization
comparative linguistics, 78, 82, 85, 90–110
competency cluster, 165–166
complementarity of signal, 162
consonant clusters, 86–87
consort relationship, 163–164
    see also pair-bonding

contact-seeking behavior, 193, 194
conviction, feelings of, 47
Conybeare, W. D., 20
Cooper, A. J., 184
Copernicus, N., 27
copulatory patterns, hormonal regulation
 of, 171–173
copulin, 182
core gender identity, 158
 see also gender identity
corpus striatum, 42
cortex, evolution of, 45–46
 and language ability, 291–292
 speech areas of, 61
corticospinal tract, 60, 67
cortisol, 174
courtship behavior, 44, 45
 hormonal effects on, 146–147
 and threat display, 163
cranial growth, primate comparison,
 268–280
creationism, 13, 19–22, 27
 and geological science, 15–16
 and natural theology, 14–15
cross-cultural studies, of gender role, 137,
 146, 156–157, 164–166
 of menstrual reactions, 179–180
 of social approach, 197
 of spatial ability, 192
culture, and biology, 287–289
 and human evolution, 77
 and menstrual behavior changes,
 179–180
 taxonomic classification and,
 216–217, 234–235
 see also cross-cultural studies
curiosity, sexual dimorphism of, 133
cultural anthropology, 288
Current Anthrolopology, 288
Cuvier, G., 12, 13, 14, 18, 20, 22
cyclicity, in the female, 169, 170
 of mood and menstruation, 177–184
 of mood and oral contraceptives,
 180–181

Dalton, K., 178, 179
Damon, A., 161
D'Andrade, R. G., 164–165

Darwin, C., concepts of, 26–27
 Galapagos visit of, 18
 influenced by Lyell, 19, 21–22
 Origin of Species, 13, 21, 23, 25–26,
 28
 species concept of, 24–25
Darwin, Marx, Wagner (Barzun), 25–26
Darwin and the Darwinian Revolution
 (Himmelfarb), 25–26
Darwin, E., 13
Darwinian revolution, 1–2, 11–12,
 25–28
Darwinism, 1–2, 27–28
 and Piltdown discoveries, 246
Dart, R., 250
Davenport, W., 131–132
Davidson, W., 191–192
dawn stones, 240, 247
Dawson, C., 241, 245, 247, 248–249,
 250
Dayang, 219, 222–224
deceptive behavior, 44, 50
deChardin, T., 27, 241, 244, 245–246,
 247
defecation posts, 44
Demos, R., 47
dentition, 274–278
dependency, 196–197
 concept of, 193
 and sex difference, 131
depression, and menstrual cycle, 177–180
 and oral contraceptives, 180–181
Descartes, R., 35
development, tempo of, 279–280
 see also ontogenesis
DeVore, I., 163, 186, 187, 194
dexamethasone, 184
dichotomous categorization, and
 sexuality, 139
Diderot, D., 13
disease, and menstrual cycle, 178
 theories of, 290–291
display behavior, 43–44, 45
Dobzhansky, T., 156
Doering, C. H., 186
Dolhinow, P., 186–187
dominance, 162
 gender variations in, 164

loss of and testosterone level, 184
see also aggressiveness, heirarchy
dopamine, 42
Dravidian language, 84, 87–90
    cognate sets, 91–110
    shared isoglosses, 115
drives, 50
Dryopithecus, 240
Dubois, E., 240
Durham, N., 233
Durkheim, E., 7, 289

East Indian Archipelago, 18–19
economic specialization, 140–144
Economists at Bay (Lekachman), 290
Ehrhardt, A. A., 171, 175–176, 186
Eibl-Eibesfeldt, I., 188
eidos concept
Eimas, P. D., 194
Einstein, A., 12, 27, 37
Eisenberg, R. A., 194
ejaculatory ability, 127, 135
Elliot, R., 240
embryo, brain development of, 16
    human and monkey, 257–262
    sexual development in, 126
Emerson, P. E., 194–195
Emerson, R. W., 13
emotion, brain centers of, 47, 50–53
    communication of, 291–292
    expression of in animals, 63–66
    and facial expression, 59–61, 68–69
    gender and expression of, 128, 164
    gonadal hormones and, 180
    "major," 51
empathetic behavior, 48, 193
endocrine environment, and brain
    development, 126–127
environment, changes in and mutation,
    166
    and evolution, 158–159
eoliths, 240, 247
epigenesis, 125
    of male and female, 125–128
    of masculine and feminine, 128–137
epilepsy and brain function studies, 47
epistemics, 36
epistemology, 35–36

Epstein, S., 187–188
Essay on Classification (Agassiz), 15, 23
essentialism, 14, 17–19, 24, 27, 28
estradiol, 179–180
estrogen, in males, 174–176, 185
    in menstrual cycle, 176–177, 179
    and sex differentiation, 128,
        169–170
    and sexual attraction, 145–146,
        181–182
estrus, 145–147
    and sexual receptivity, 181–182
    and stable pair-bonding, 163–164
Ethnographic Atlas, 165
ethnography, 288
ethnology, 290
ethnosemantics, 116
ethology, 40, 43, 161, 286
    and primate taxonomy, 216–217,
        234–235
Ever-Ready Vagina (ERV) theory,
        144–145, 149
Everitt, B. J., 184
evolution, asymmetric theory of,
        242–243, 249–250
    of gender roles, 138–150
    human and behavioral, 286
    of sexuality, 124
    and social science, 294
    of social systems, 140–150
    of speech, 63
    synthetic theory of, 159–160
Evolution of Behavior (Brown), 286
evolutionary anthropology, 27
evolutionism, 21
    pre-Darwin concepts of, 12–13
expression, facial, 59–64
    see also emotion
extinction, and creationism, 16, 20
    and speciation, 24

facial expression, 59–64
facts, 38–39
Falck, B., 42–43
family structure, primate and human,
        142–150
    recent changes in, 157
fear, 51

feeding, and olfactory sense, 48
female, dependency concepts and,
    196–197
    economic roles of, 140–144, 157,
        164–165
    gender attributes of, 131–137
    intimidation of, 164–166
    occupation of, 161
    orgasm in, 146–149, 163
    in preagricultural society, 140–144
    verbal skills of, 189–191
    see also male-female dyad, women
female hormones, and sexual behavior,
    172–173
    see also estrogen, progesterone
feminine gender role, and evolution,
    138–150
    see also sex role
femininity, 156
    attributes of, 131–137
    behavioral studies of, 172
    see also masculine-feminine dyad
Ferrier, D., 41
Ferris, B. G., 161
Feshbach, S., 186
fetalization theory, 257
Field, P. M., 171
fighting, territorial, 44
    see also aggressiveness, territoriality
flaked tools, see toolmaking
flocking behavior, 44
    see also affiliative behavior, social
        bonding behavior
flood, see catastrophism
fluorine, in fossil bones, 243
follicular phase, 176–177, 179, 181
food sharing behavior, 140, 141
foot, ontogenic changes in, 258–262
foraging behavior, 44
forebrain, 40–41
Forel, A. H., 35
Foss, G. L., 183
Fossey, D., 187
fossils, dating of, 243
    and extinction, 24
    see also eoliths
Foster, M. L., 77–118

Frank, R. T., 178
Frankenhaeuser, M., 184
Freedman, D., 185
Freud, S., 35, 40, 50
fricatives, 90
    +*e reflexes, 89
    syllabic allophones of, 88
From Instinct to Identity (Breger), 286
frontal lobes, 60–61

Galapagos Islands, 18
games, see play
gathering activities, 164
Gaynor, F., 78
gender constancy, 130
gender identity, 128, 137
gender role, 128–137
    adaptive function of, 143
    see also sex role
generosity as social value, 141
    see also food sharing
genes, 159–160
    and reproductive success, 166–167
    see also chromosomes
genetic code, 159
genetic variation, 17
genital function, and olfactory sense, 48
genotype, 158
geographic distribution, 16
geology, and evolutionism, 16
gestures, 291
    see also nonverbal behavior,
        prosematic behavior
gibbon, 163, 201, 271, 272, 280–281
Gillen, F. J., 161
Glass, G. S., 178, 179
globus pallidus, 42
Goldberg, S., 190
gonadal hormones, 169, 171–173
    and sexual activity, 181–185
    see also hormones
gonadal primordium, 169
gonadotropin, 169–173, 185
Goodall, J. V., 187, 192, 194, 216, 218
gorilla, 179, 261, 267–269, 270–271,
    276
Gorski, R., 170, 171

Gottschalk, L., 178
Graafian follicle, 176–177
Grange, K. M., 35, 51
Grant, E. C., 182
Gray, J. A., 189
gray matter, 42
Green, H. D., 67
Green, R., 171
Greene, R., 178
greeting behavior, 44, 45
grip strength, 161
grooming behavior, 44
    and menstrual cycle, 181
Grounds, A. D., 182
group living forms, and sexuality,
    139–150
growth, tempo of, 279–280
guerezas, 260
Guthrie, R. D., 162–163, 188

habitus, female, 127
Haeckel, E. H., 240
Hahn, J. D., 195
hair, facial, 127
Halasz, B., 171
Hall, K. R., 186, 187
hamadryas baboon, 149, 163, 183
Hamburg, B. A., 155–198
Hamburg, D. A., 180
Hammond, W. A., 133–134
hand, ontogenic changes in, 258–263
handedness, 188–189
Hansman, C., 191
Harlow, H., 173
Harris, G. W., 171
Hayes, G., 72
Hayes, K. J., 72
Haynes, H., 194
head size, 263, 265
heart rate, 161
heirarchy, and language, 116, 118
    in nonhuman primates, 163
    social, 44, 162
Heisenberg, W., 12, 27
heleolithic wave, 249
Helmholz, H. L., 38
help-seeking behavior, 193

hemispheric specialization, 292
    see also brain, handedness
hemoglobin level, 127, 134
Herbert, J., 184
hermaphroditism, progestin induced,
    174–175
    in rhesus monkeys, 173–174
    see also pseudohermaphroditism
Herrick, C. J., 49, 246
Herrmann, J. B., 183
Herschel, J. F., 21
heterosexual attraction, 135–137
heterozygosity, and mutation, 166–169
    and x-linked diseases, 168–169
Hillarp, N. A., 42–43
Hinde, R. A., 188, 193–194, 197
Hittite language, 82, 84–85, 87–90,
    91–110
hoarding behavior, 44
Hockett, C. F., 63
holistic perception, 191
Holloway, R. L., 216
homesite, selection and preparation, 43
homing behavior, 44
hominids, affectional bonds among, 163
    brain development in, 161
    early finds, 240
    foot development of, 261
    imitative learning by, 218–235
    span of existence of, 160
    specialization by, 267–269
    taxonomy of, 216–217, 234–235
Homo erectus, 217
Homo habilis, 217–218
Homo sapiens, body proportions of, 267,
    269
    evolution of sexuality in, 124–125
    family structure of, 142–150
    pair-bonding among, 162
    span of existence of, 160
    speech areas of, 68
    toe development of, 261
Hooker, J. D., 14, 22
Hooten, E. A., 242
hormones, cross-sex presence of, 185
    differential effects of, 180
    female, and sexual activity, 145–146

hormones (*continued*)
    and neuroendocrine function,
      170–171
    and nurturant behavior, 195
    ovarian, 135
    and sexual behavior, 135–137,
      171–173, 181–182
    and sexual development, 125
      169–170, 174–176
    see also oral contraceptives,
      progesterone, progestin,
      testosterone
hostility, and facial expression, 63–64
howler monkey, 257
human, as scientific term, 234–235
    see also Homo sapiens
Hume, D., 38
hunting, 44
    and communication, 161–162
hunting-gathering society, 160–161
    and sex role, 139–150
    social organization of, 163
Hutt, S. J., 194
Hutton, J., 19, 20
hypersexual behavior, 184
hypothalamus, 46, 49, 61
    arcuate (tuberal) region of, 171
    sexual dimorphism and, 126–127,
      170

id, 50
idée fixe, 52
Iliff, A., 161
imitative learning
    by chimpanzees, 218–219
    by orangutans, 219–235
    see also isopraxic behavior, learning,
      social learning
independence, and attachment, 197
individuals, physical characteristics of,
    159
Indo-European language, 84–89
infancy, 129
    attachment behavior in, 194–195
    dominance behavior in, 195
innominata area, 61
Inskeep, R. R., 217

instincts, 50
    and facial expression, 63
    and vocalization, 68–69
intellectual domain, 51–52
    see also territoriality
intelligence, 287
    and affect, 47
    sexual dimorphism in, 133, 190
Interpretation of Dreams (Freud), 35
IQ, sexual dimorphism in, 133, 190
Ismail, A. A., 183, 184
isoglosses, 114, 115
isopraxic behavior, 44–45, 50
    see also imitative learning, learning

Jacklin, C., 186, 187, 189, 190, 193,
    197
Jacobs, T. J., 179
Jacobsohn, D., 171
Janiger, O., 179
Janowsky, D. S., 179
Jasper, H., 68
Java man, 234
Jeans, J., 53
Jensen, G. D., 196
Johnson, V. E., 146–149
Jost, A., 169
Jurgens, U., 67

Kagan, J., 194
Kane, F. R., 182
Kant, E., 53
Kay, P., 82
Keith, A., 5–6, 241, 242, 244, 246, 247
Kepler, J., 52
Keverne, E. B., 183
Kimura, D., 189–190, 191
Klinefelter syndrome, 167–168
knowledge, and culture, 35
Knowles, E., 178
Knox, C., 189, 190
Kolodny, R. C., 184
Kopell, B. S., 178
Korner, A., 195
Kramp, J. L., 179
Kreuz, L. E., 184, 186
Krogman, W. M., 239–251

Kuhn, T. S., 12
Kummer, H., 163, 183
Kutner, S. J., 180

Laboratory of Brain Evolution and
    Behavior, 44–45
lactation, 128
lactic acid neutralization, 161
ladder of perfection, 15–16, 23–24, 27
Lamarck, J. B., 12, 13, 17–18, 21,
    23–24, 26–27, 28
Lancaster, C. S., 142–143, 160–161
Lancaster, J. B., 63, 291
language, acquisition of and gender
    identity, 129–130
    changes in, 86
    cognate forms of, 91–110
    evolution of, 3–4, 77–78, 161
    origins of, 3–4, 292
    reconstruction process, 80–91
    and social science, 291–293
    see also speech, verbal learning,
        vocalization
langur, 260
Lankester, R., 241, 247
Larson, C., 68
laryngeal hypothesis, 90
Laschet, L., 184
Laschet, U., 184
Lavoisier, A. L., 12
Leakey, M., 217–218
learning, biological factors of, 159
    imitative, 218–235
    and R-complex, 45
    and rate of maturity, 162
    sexual differences in, 158–159
    see also imitative learning, isopraxic
        behavior, social learning
Lee, V. A., 161
left cerebral dominance, 242
Lekachman, R., 290
LeMagnen, J., 183
Leutenegger, W., 263
LeVine, R., 165
Levy, J., 189, 191, 192
Lewis, M., 190, 194
lexeme, 116

libido, and maternalism, 195
    and testosterone levels, 183–185
life sciences, 35
limbic lobe, 45–46
limbic system, and emotions, 46–48
    subdivisions of, 48–49
    vocalization and, 67, 291–292
Lindeman, R. C., 68
linguistics, 292
    see also comparative linguistics,
        language
Linnaeus, 24
Lisk, R. D., 171
Livingstone, F. B., 162–163, 185
lizards, as experimental animals, 43
longevity, and sex, 157
Loraine, J. A., 184
Lorenz, K., 40
Lovejoy, A. O., 12–13, 21, 22
Lunde, D. T., 180
luteal phase, 176–178, 179, 181
Luttge, W. G., 173, 181
Lyell, C., 12, 13, 14, 15
    anti-evolutionary arguments of,
        22–25
    impact on Darwin, 19–22
    species concept of, 16–22
    steady-state concept of, 28
    and uniformitarianism, 19–22, 28
Lysenko, T. D., 291

macaques, 264–265, 279
Maccoby, E. E., 186, 187, 189, 190,
    193, 197
MacLean, P. D., 33–54, 289
males, competition and cooperation
        among, 162, 194
    economic roles of, 140–144, 161,
        164–165
    gender attributes of, 131–137
    role strain in, 157
    see also masculinity
male-female dyad, 124
    epigenesis of, 125–128
    and human evolution, 141
Mall, M. G., 179
mammalian brain, 42, 45–47, 170

Mandell, A., 178
Mandell, M., 178
mandrill, 273, 275
Maresh, 191
marking of territory, 43
marmoset, 277–278
Martinez, C., 171
masculine-feminine dyad, 124
    epigenesis of, 128–137
masculine gender role, and evolution,
        138–150
    see also sex role
masculinity, 156
    behavioral studies of, 172
    in hermaphroditic girls, 175–176
Masters, J. C., 193
Masters, W. H., 146–149
maternal behavior, and pair-bonding, 196
    see also mother-child bond
mathematical aptitude, 133
mating behavior, 44
    and brain development, 126–127
    and olfactory sense, 48
mating position, and human sexuality,
        148–149
maturation, 129
    differential rates of, 162, 190–191
Maupertuis, P. L., 12
Mayr, E., 11–28
McCammon, R. W., 191
McCance, R. A., 276
McCarthy, D., 189
McClintock, M. K., 182
McCown, E. R., 192–193, 285–294
McGlone, J., 191–192
Mead, A., 140, 143–144
meaning, and articulation, 110–114
    and comparative linguistics, 86
    and position in language, 117
Mears, E., 182
medical anthropology, 286, 290–291
medulla, 41
Megaladapis, 267–268, 273
Melanesian society, sex roles in, 131–132
menarche, 126, 135
menstrual cycle, cross-cultural
        observations of, 179–180
    endocrinology of, 176–179

onset of, 126, 127, 135
    and pair-bonding, 163–164
    synchrony of, 182–183
menstrual phase, 176, 179, 181
mentation, forms of, 49–50
metaphor, and meaning, 118
Meyer-Bahlburg, H., 186
Michael, R. P., 181, 182, 183
microcatastrophism, 18
    see also catastrophism, 273
Microcebus, 273
midbrain, 41, 68
migration behavior, 44
Millar, R., 239, 244, 247, 250
Miller, G., 247
Miller, G. S., Jr., 242
Milner, B., 189, 192
Mitchell, G. D., 195
Money, J., 158, 170, 175, 184, 186
modern world, vs. primitive world,
        288–290
Moir, J. R., 247
monkeys, experiments with, 45, 49
    longtailed, 258
    sexual dimorphism in, 274
    vocalization and expression in,
        63–71
    see also primates
Monod, J., 37
monogenesis of language, 81–82
Moore, T., 190
Moos, R. H., 178
morphemes, 82, 84, 110
Morris, N. M., 181
Morsbach, H., 39
Moss, H. A., 190
mother-child bond, 162, 194–196
    and verbal abilities, 190
mountain gorilla, 261
Moyer, K. E., 187–188
Muller, H. J., 25
Mullerian ducts, 125, 169
Murchison, R. I., 20
muscle and bone development, sexual
        dimorphism in, 127
    and testosterone, 161
mutation, and evolution, 159
    preservation of, 140

reproductive effects of, 166
Myers, R. E., 59–73, 291
Myers, S., 65, 68

Naftolin, F., 185
Nagylaki, T., 192
Napier, J. R., 230–231
National Institute of Mental Health, 45
natural sciences, 35
natural selection, *see* selection
natural theology, 14–15
Neanderthal man, 240
Nebes, R. D., 191
nesting behavior, 43
neocortex, 46, 49–50
neomammalian brain, 36–37, 41, 46, 48,
    53
Neuman, F., 195
neuroendocrine function and sex
    hormones, 170–171
New Introductory Lectures (Freud), 50
Newton, I., 12, 27
Newton, N., 163, 195
Nielsen, J. M., 62
nomialism, 27
nonamygdaloid dendritic spine synapses,
    171
nonverbal behavior, 39–41, 291
nuclear family, *see* family structure
nurturant behavior, 44, 140, 163–164,
    193–194

Oakley, K. P., 241, 243
obsessive-compulsive behavior, 50
Oetzel, R. M., 186
Olduvai site, 217
olfactory function, 48
    and sexual behavior, 182–183
olfactostriatum, 42
ontogenesis, 125
    psychosexual, 129, 135–137
    recapitulation theory of, 256–262
opinion, 51–52
opposition, and language, 116, 118
oral contraception, 182
    and cyclic mood change, 180–181
oral tract, and language, 111
    see also articulation

orangutan, comparative anatomy of, 261,
    264, 266, 268, 269, 270, 276,
    280–281
    flaked stone toolmaking by,
        219–235
    hand structure of, 231–233
    mental capacities of, 233–234
orbitofrontal region, 60
Origin of Species by Means of Natural Selection
    (Darwin), 13, 21, 23, 25–26,
    28
orosexual manifestation, 48
Ossemens Fossiles (Cuvier), 13
ovaries, 125, 128, 169
ovarian hormones, 176–178
    see also estrogen, progesterone
ovariectomy, and hormone studies,
    181–182
ovulation, 176, 178, 183
    and mood change, 180
    spontaneous, 171
Owen, R., 13, 14

Paige, K., 180
pair-bonding, origins in Homo sapiens,
    142–150, 194, 196
    and sexual competition, 162–164
paleoanthropology, 161
paleomammalian brain, 36–37, 41,
    45–49, 53
paleopsychic processes, 38, 40–41,
    52–53
Paley, W., 14
Pallie, W., 189
Pan paniscus, 278
Pan troglodytes, 278
paradigmatic linguistic description, 83,
    116–117
parental behavior, 44
    see also mother-child bond,
        paternalism
Parlee, M. B., 179
parturition, 128
passive-dependent concepts, 196–197
paternalism, 195–196
Patterson, G. R., 187
Pederson, F. A., 193
pedomorphism, 162

Pei, M., 78
pelvis development, 263
Penfield, W., 62, 68
penile bone, 145
penis, 126
periaqueductal gray substance, 68
perpetual intervention hypothesis, 20
perseverative behavior, 44, 133
Persky, H., 179, 183, 186
personality, 49
Pfeiffer, C. A., 170
phememes, 78–80, 114
    comparison of, 90–110
phenotype, 158
Philosophie Zoologique (Lemarck), 28
Phoenix, C. H., 173, 186
phonemes, 78, 81, 84, 292
phonetics, 292
phonological isoglosses, 114–115
phylogeny, and recapitulation theory,
    256–262
physical anthropology, 286–287
Piaget, J., 47
pill, the, *see* oral contraceptives
Piltdown man, 5–6, 239–251
    Piltdown I, 241–245, 247
    Piltdown II, 244–245
Piltdown Men, The (Millar), 239
Pinel, P., 51
Pithecanthropus, 240
pituitary glands, 170–173
pituitary-ovarian feedback system,
    176–177
place, preference for, 43
Planck, M., 12, 52
Plant, J. M., 182
Plato, 17, 47
Platyrrhines, 260, 274
play, and attachment behavior, 194
    cross-cultural studies of, 197
    sexual dimorphism in, 131–133,
        173–174, 175–176, 186–187
Ploog, D., 291
Poe, E. A., 219
Pokorny, J., 85
pongids, 234, 264, 266
pons, 41, 68

Popper, K. R., 17, 22
population, doubling rate of, 34
population theory, 19, 27
    and gender role variation, 138–139
    and human behavior, 159
    and sex differences, 127, 132
Porteus, S. D., 192
postindustrial society, 160
posture, primate comparison, 267–269
    see also bipedalism
potto, 260, 262
preadaptation, 166
preagricultural society, and evolution of
    sexuality, 139–150
precedent, obeisance to, 50–52
precentral gyrus, 60, 67, 68–69
precision grip, 231
prefrontal cortex, 48
prefrontal-orbitofrontal cortex, 70–71
pregnancy, 128
    and social role, 140
    stress during, 172
premenstrual phase, 176–179
premenstrual syndrome, 177–179
preoptic area, and ovulation, 171
primary processes, 40
primates, agressiveness among, 162,
        185–188
    attachment behavior in, 196
    characteristics of, 138, 140–141
    dental abnormalities in, 277
    digital development of, 260–263
    hierarchical orders among, 163, 165
    mother-young subgroups among,
        195
    paternalism among, 195
    phylogenetic specialization among,
        255–280
    sexual cycles among, 145
    sexual activity and pair bonding
        among, 149–150
    simian body proportions, 263–267
    span of existence of, 160
    vocal and facial function among,
        66–72, 291
    see also Homo sapiens, monkeys
primatology, 286

primitive world, 288–290
primordial language (PL), 78, 80, 83–84,
    87
    cognate sets, 91–110
    conceptualization of, 111
    shared isoglosses, 115
Principles of Geology (Lyell), 15, 20, 21
problem-solving behavior, 133
proboscis monkey, 273, 276
procreation, brain centers of, 46–47
    see also sexual behavior
progesterone, levels in menstrual cycle,
    177–178, 179
    and sexual interest, 181, 182
progestin, hermaphroditism induced by,
    174–175
    and sex drive, 183–184
progressionists, 14, 19, 25, 26–27
    see also catastrophism
Propithecus, 267–268
prosematic behavior, 38–41, 45, 49–50
    see also nonverbal behavior
prosimians, 267
Proto-Indo-European (PIE), 82, 83–84,
    87–90
    cognate sets, 91–110
    shared isoglosses, 115
protomentation, 49–50
protoreptilian propensities, 51–52
    see also R-complex, reptilian type
        brain
proximity-seeking behavior, 193, 194
pseudohermaphroditism, 173–174
psychiatry, 35
psychoanalysis, 35
psychology, 34, 35, 38
psychotherapeutic drugs, 47
puberty, *see* adolescence
Puhvel, J., 82
putamen, 42
Pycraft, W. P., 241, 247

quadrupedalism, 267–269

R-complex, 42–45, 47
    see also reptilian type brain
Raisman, G., 171

Ramapithecus, 240
rational mentation, 50, 52–53
Ray, J., 26
reason and emotion, 47
recapitulation theory, 256–262
Reddy, V. V., 185
reenactment behavior, 44
Rees, O., 243
Reiser, O. L., 17
relationship patterns, 193–197
reptiles, mammallike, 45
reptilian type brain, 36–37, 41, 42–45,
    49, 53
reproduction and evolution, 159
reproductive anatomy, 126, 127
reproductive behavior, 166–173
    see also sexual behavior
resonants, syllabic allophones of, 89
responsibility, gender variations in, 164
rhesus monkeys, 2
    experimental hermaphroditism in,
        173–174
    fetal studies of, 258, 264–265
    menstruation and mood among, 179
    sex hormones and behavior of,
        181–182
    vocalization by, 63, 65–66, 67–71
Ribeiro, A. L., 178, 179
ritualistic behavior, 50
Rivarola, M. A., 185
Roberts, L., 62
Robinson, B. W., 67, 291
Robinson, J. T., 217, 234
Rock, J., 178
romantic love, 136–137
Rose, R. M., 184, 186
Rosenblatt, J. S., 173, 195
Rosenblum, L., 173, 186, 196
Rossi, A., 179, 180, 183
routinization of behavior, 44
Rubenstein, B., 177–178

Sahlins, M., 141–142
St. Hilaire, E. G., 13
Sanday, P. R., 165
Saunders, F. G., 170
Saussure, F. de, 83

scala naturae, 15–16, 23–24, 27
Schaffer, H. R., 194–195
Schaller, G. B., 219
schizophysiology, 47
Schopenhauer, A., 13
Schultz, A. H., 6–7, 255–281
Schwartz, N., 176
scientific aptitude, 133
scientific methodology, and evolutionism, 22–24
scientific revolution, 11–12, 25
scrotum, 126
secondary sexual characteristics, 162
sedentism and gender, 131
Sedgwick, A., 20
selection, 19, 27–28
    and sex differences, 138, 155–156
    and signalling ability, 162
self, science of, 36
self-concept, and sexuality, 128–129
self-preservation, behaviors, 43–45
    brain centers of, 46–47
semantic common denominators, 91–110
semantics, and articulation, 80
    shifts in, 83
sensory systems, and brain function, 49
    and sexual development, 134
sex, as biosocial problem, 287
sex attractants, 182–183
sex hormones, and behavior, 181–182
    see also hormones
sex linkage, 168–169
sex role, 157
    -adoption, 158
    cross cultural studies of, 164–166
    stereotypes, 156–157, 165–166
    -orientation, 158
    -preference, 158
    see also gender role
sexual behavior, brain regulation of, 171–173
    gender variations in, 164
    pheromones and, 182–183
    sex hormones and, 181–182
    testosterone and, 183–184
    see also reproductive behavior
sexual dimorphism, 124–128, 274–277
    in affiliative behavior, 193–197

in aggressiveness, 185–188
in brain, 171–173, 188–193
chromosomal, 167–169
hormonal, 169–170
in language ability, 133, 189–191
in learning ability, 158–159
in spatial perception, 191–193
sexuality, and adolescent development, 135–137
    epigenesis of, 124, 128–134
    evolution of, 4–5
sharing behavior, 140, 141
Sherman, J. A., 190
siamang, 272
sign, and language, 83, 114, 118
signalling ability and selection, 162
simian primates, 263–267, 273
Sinanthropus, 246
Skinner, B. F., 64
skull development, primate comparison, 268–280
Smith, C. W., 161
Smith, G. A., 241, 242, 246, 247, 249–250
Snow, C. P., 37
social anthropology, 285
social bonding behavior, 44, 138–144, 193–197, 287
social emotional behavior, 70–71
social facilitation, 44
    see also isopraxic behavior, learning
social learning, and agression, 186–187
    and dependency behaviors, 193–194
    of gender role, 128–137
    of sexual behavior, 173
    see also learning
social organization, 160, 163
    efficiency of, 293–294
social science, 35, 286
    effectiveness of 293–294
sociobiology, 285–294
Sociobiology (Wilson), 7, 286, 287
Sommer, B., 179
space, concepts of, 53
spatial perception, 161
    sexual dimorphism in, 133, 167–168, 191–193
species, adaptability of, 24–25

historic theories of, 16–22
variations within, 278–279
species-specific behavior, 40, 138
species-typical behavior, 40, 45,
simian, 49
speech, and brain function, 60–62
and hemispheric dominance,
189–191
see also language, verbal behavior,
vocalization
speech cortex, 61
Spencer, B., 161
Spencer, H., 13, 38
Sperry, R. W., 189, 191, 192
spider monkey, 260, 278
spinal cord, 41
squirrel monkey, 45
Starck, D., 257
status, and signalling ability, 162
steady-state world, 24, 26–28
Steinbeck, H., 195
Sterkfontein site, 217
steroids, *see* hormones
Steward, J. H., 161
stress in pregnancy, 172
subjectivity, 38
submission signals, 44, 188
subthalamic region, 49
subthalamus, 61
superstitious behavior, 50
surrender, signals of, 44, 188
survival, behaviors involving, 43–45
sex role and, 143
Sussex Archaeological Society, 241
Sutton, D. C., 66, 68
Swadesh, M., 82, 114
Swartkrans site, 217
symbol and language, 83
symbolic behavior, and brain function,
49, 62
evolution of, 77–78, 116–118

tail, ontogenetic development of,
257–258
Talgai skull, 250
Tanner, J. M., 161, 191
Tarascan language, 91–110, 114
Tarsius, 267

taxonomy, and toolmaking ability,
216–217, 234–235
Taylor, D. C., 190
Taylor, S. P., 187–188
teeth, growth of, 274–278
temporal lobe, 61
territoriality, 43
and intellect, 51–52
and spatial ability, 192
testes, 125, 169
testicular hormone, 127
see also testosterone
testosterone, 126
and aggressiveness, 185–186
anabolic effects of, 161
and brain development, 167
in females, 174–176, 183–185
in homosexuals, 184
and male sexual behavior, 184–185
morphology and human behavior,
172
mother-child interactions and, 196
and pseudohermaphroditism,
173–174
psychosocial effects in males,
184–185
secretion rates of, 127, 135, 136
and sex differentiation, 169–170
thalamus, 61
Thomism, 17
thought, and emotion, 47
threat, and signalling ability, 162–163
thumb, abducted, 231
time, concepts of, 53
Tinbergen, N., 40
Tobias, P. V., 218
toes, development of, 260–262
Tonge, C. H., 276
toolmaking, 5, 80, 160, 161
Australopithecines and, 217–218
behavioral significance of, 216–217
by and orangutan, 219–235
and sex role, 164
touch pads, 259–260
Tournefort, J. P., 26
trail making behavior, 43
tree shrew, 266–267
triune brain, 41

tropistic behavior, 44
Turiaf, J., 178
Turkish language, 84–85, 87–90
    cognate sets, 91–110
    shared isoglosses, 115
Turner's syndrome, 167
Tyler, L., 189

Udry, J. R., 181
Unger, F., 13
uniformitarianism, 19–22, 28
Unready Penis (URP) theory, 144–145
urogenital sinus, 126
uterus, 126

vagina, 126
    and sexual satisfaction, 146–149
    see also Ever-Ready Vagina theory
Valenstein, E. S., 173
vanLawick-Goodall, J., see Goodall, J. V.
Vandenbroek, G., 278
variation, 19
venturesomeness, 131
verbal behavior, 38, 41
verbal learning, and gender identity,
    129–130
    sexual dimorphism in, 133,
    189–191
Vestiges of the Natural History of Creation
    (Chambers), 13, 15, 22
Vierling, J. S., 178
visual function, 48
vocalization, 60, 291–292
    in animals, 62–63
    and expression, 61–62
    and facial expression, 63–66
voice pitch changes, 127

Wade, N., 287
Walker, A. E., 67
Wallace, A. R., 13, 19, 21
Ward, I., 172, 185
Washburn, S. L., 52, 142–143,
    160–161, 163, 192–193,
    285–294
Watson, H., 13–14
Watson, J. B., 38
Waxenberg, S. E., 183

Weiner, T. S., 243, 244
Weismann, A., 28
Wells, L. H., 216, 234
Wetzel, R. D., 178
Whalen, R., 170, 171, 185
Whewell, W., 20
Wiener, N., 38
Wilson, E. O., 7, 286
Wilson, L. G., 19
Wilson, W. C., 187
Witelson, S. F., 189
Witkin, H. A., 192
Wolffian ducts, 125, 174
women, changing social roles of, 157,
    164–165
    cyclic mood changes and
        contraception, 180–181
    sexual behavior and testosterone,
        183–184
    social roles and menstruation,
        179–180
    subordination of, 164–166
    see also female, femininity, Homo
        sapiens, male-female dyad,
        masculine-feminine dyad
Woodward, A. S., 240, 241, 244, 245,
    247
Woolsey, C. N., 68
Wright, R. V. S., 215–235

X chromosomes, 125
    conditions linked with, 168–169
    and sex differentiation, 167–168
XX fetus, 126
XY genotype, 126, 143

Y chromosomes, 125
    and sex differentiation, 167–168
Yalom, I., 176
Yamaguchi, S., 65, 68
Yarrow, L. J., 193
Yerkes Primate Center, 179
Young, W. C., 173
Yurok language, 84–85, 87–90
    cognate sets, 91–110
    shared isoglosses, 115

Zinjanthropus, 217–218
Zuckerman, S., 140, 143–144

14
1